THE ORIGIN OF MAN

THE ORIGIN OF MAN

BY

CARVETH READ, M.A. (CANTAB.)

EMERITUS PROFESSOR IN THE UNIVERSITY
OF LONDON

SECOND EDITION
REVISED AND
ENLARGED

CAMBRIDGE
AT THE UNIVERSITY PRESS
1925

CAMBRIDGE UNIVERSITY PRESS
Cambridge, New York, Melbourne, Madrid, Cape Town,
Singapore, São Paulo, Delhi, Tokyo, Mexico City

Cambridge University Press
The Edinburgh Building, Cambridge CB2 8RU, UK

Published in the United States of America by Cambridge University Press, New York

www.cambridge.org
Information on this title: www.cambridge.org/9781107628007

First edition 1920
Second edition 1925
First paperback edition 2011

A catalogue record for this publication is available from the British Library

ISBN 978-1-107-62800-7 Paperback

PREFACE TO *THE ORIGIN OF MAN AND OF HIS SUPERSTITIONS*

FIRST EDITION

THE volume now published explains in its first part an hypothesis that the human race has descended from some ape-like stock by a series of changes which began and, until recently, were maintained by the practice of hunting in pack for animal food, instead of being content with the fruits and other nutritious products of the tropical forest. The hypothesis occurred to me many years ago, and was first published (in brief) in *The Metaphysics of Nature* (1905), Chap. xv, § 3, and again in *Natural and Social Morals* (1909), Chap. vii, § 2; but all it implied did not become clear until, in lecturing on Comparative Psychology, there was forced upon me the necessity of effecting an intelligible transition from the animal to the human mind, and of not being satisfied to say year after year that hands and brains were plainly so useful that they must have been developed by natural selection. Then one day the requisite ideas came to light; and an outline of the hypothesis was read at a meeting of the British Association (Section H) at Birmingham in 1913, and printed in *Man*, November 1914. The Council of the Anthropological Institute has kindly consented to my using the substance of that article in the first chapter here following.

The article in *Man* dealt chiefly with the physical changes which our race has undergone. The correlative mental changes were explained in the *British Journal of Psychology* in an article which supplies the basis of the second chapter of this book.

The hunting-pack, then, was the first form of human society; and in lecturing on Ethnopsychology two questions especially interested me: (1) Under what mental conditions did the change take place from the organisation of the hunting-pack (when this weakened) to the settled life of the tribe or group? and (2) Why is the human mind everywhere befogged with ideas of Magic and Animism? They seemed at last to have the same answer: these superstitions were useful and (apparently) even necessary in giving to elders enough prestige to preserve tradition and custom when the leader of the hunt was no longer conspicuous in authority. A magic-working gerontocracy was the second form of society; and the third form was governed by a wizard-king or a priest-king, or by a king supported by wizards or priests. One must, therefore, understand the possibility of these beliefs in Magic and Animism, and how they arose and obtained a hold upon all tribes and nations; and hence the second part of this volume—on Superstition.

Some results of inquiry into these matters were also published in the *British Journal of Psychology* (namely, much of the substance of Chaps. III, IV, V, VI, and VIII) and are here reproduced, with the editor's consent, enlarged and, for the most part, rewritten: the least altered are Chaps. VI and VIII. Chaps. VII, IX and X have not hitherto been printed; but part of Chap. X was read at the Meeting of the British Association at Bournemouth last year.

Messrs Williams and Norgate have given permission to use the diagram in the footnote to p. 3, based on one of Professor Keith's in his *Antiquity of Man*.

Extensive use has, of course, been made of the works of Darwin, Herbert Spencer and E. B. Tylor, and (among living authors) of the volumes of Sir J. G. Frazer and Prof. Ed. Westermarck. I am grateful to my friends and colleagues, Prof. Spearman, Prof. J. P. Hill and Prof. Arthur Keith for assistance in various ways. Mr Pycraft, too, of the Natural History Museum has given me important information; and my old friend Mr Thomas Whittaker, has helped me, as usual, when my need was greatest.

C. R.

University College, London,
4 *July*, 1920.

PREFACE TO *THE ORIGIN OF MAN*

SECOND EDITION

THE original volume on *The Origin of Man and of His Superstitions* appeared to some reviewers and to some of my friends to consist of two parts not closely enough connected to justify their inclusion in one book. My own view was that Man as we know him lives everywhere under some kind of government or social regulation, which again depends for its efficacy (except perhaps in the most civilised states) very much upon the prevalence of certain superstitious beliefs. Social life is more influential than anything else in developing the mental and moral qualities that constitute his true humanity. Since, then, his social life has depended on his superstitions, an account of these is necessary to the understanding of his origin: a biological explanation is not enough. Still, the reflection that many who are interested in zoological Man may feel little concern for his beliefs, whilst others to whom these beliefs are an engrossing study may care little about his physical evolution, made it seem desirable to produce the two parts in separate volumes. Accordingly, this volume presents only the first part (Chaps I and II) of the original work, rearranged and much enlarged; and another book on *Man and His Superstitions* (comprising the last seven chapters of the former work) treats of Belief in general, Magic, Animism and related subjects.

In a long footnote at the end of Chap. I, § 1 (p. 4 of the first edition) are mentioned all the works indicating agreement with my account of our remote ancestry that were known to me four years ago. Recently, however, a much more explicit statement of these ideas has come to my knowledge in *Man and His Ancestors* by Mr Ch. Morris[1]; and I take this opportunity of repairing my oversight. Mr Morris has clearly stated the hypothesis that Man was differentiated from the anthropoids by

[1] Macmillan and Co., N. York, 1900; 2nd imp. 1902.

adopting the life of a hunter, and has drawn some of the consequences. It is comforting to have a close ally in such a risky undertaking. Mr Morris is (I suppose) an American; and to give our ideas prestige it is a pity he should not have been more of a foreigner. Still, as the hypothesis must now be considered his property in virtue of prior publication, I am relieved of the obligation to be modest in speaking of it, and might even (if occasion offered) venture the length of saying how well I think of it.

Though some readers have found my argument plausible, few have been satisfied with it, and many regard it as vain and fanciful. The fact is that most students of animate nature are busy with experimental Biology, and admirable results they have obtained. It is the fashion to neglect general reasoning (though like ours proceeding upon acknowledged facts) that does not lead to direct experiment; and in this case experiments cannot be made. One reviewer thinks it enough to say—"Die Zeit, in welcher man für solcher Hypothesen in der Wissenschaft Interesse hatte oder sogar sich begeistern konnte, ist wohl endgültig vorüber." In short, it is out of fashion, and there is as much a fashion in philosophemes as in furbelows. A well-disciplined pack follows only one quarry at a time. So irresistible is the fashion that we are told that "a generation has arisen that knows not Darwin." The unhappy generation must have been badly taught. No wonder they will not listen to Mr Morris. The arrival of Man in the world is the most wonderful event since the formation of protoplasm; yet in minds that cannot explain it in their own way it excites no curiosity. Content with the evidence of their close relationship to the apes, they are indifferent to the causes that have made them a little lower than the angels.

I must venture to say that Man is not explained by showing his relationship to the apes. That is a classification on the ground of resemblance, and the resemblance is sufficiently explained by the hypothesis of heredity. But Man is constituted not by his resemblance to the apes but by his differences from them, and the differences cannot be explained by heredity. To indicate the probable cause of the differences is the purpose of Mr Morris's book and of mine. Such a problem cannot be peculiar to the case of Man: every species of plant and of animal presents the same problem; and we sometimes meet with suggestions as to the causes of their differentiation—as that the horse's single-toed hoof is an adaptation to open dry plains; that the spots, or stripes, or fulvous coloration of certain cats is an adaptation to life in forest, or jungle, or desert; and many other examples of "protective mimicry." With each species we may inquire—what causes in its genesis, habits or environment so modified its heredity as to establish its differences. The task is intimidatingly vast; but until it is accomplished in a considerable number of instances we do not understand evolution. Natural selection gives no particular explanation of anything unless we can point to the particular conditions (*a*) of variation (which may be investigated experimentally), and (*b*) of the habits and environment which determined the selection. For instance, in our own case, the environment (probably open or thinly-wooded country), wherever situated, must have been favourable to the enterprise of that remote ancestor of ours who took to hunting for a livelihood, and whose descendants, continuing that habit, acquired age by age the specific characters that constitute human nature.

C. R.

Solihull,
September, 1924.

CONTENTS

PAGE

Preface to *The Origin of Man and of His Superstitions* . . v

Preface to *The Origin of Man* (2nd edition) vii

CHAPTER I

An Hypothesis concerning our Origin 1

§ 1. Man was differentiated from the Anthropoids by becoming a hunter; perhaps in the Oligocene period. Difficulties in beginning the adventure. Its locality 1–6

§ 2. Natural Selection; how understood in this book 7–8

§ 3. Logical character of our hypothesis 8–10

CHAPTER II

Prey and Competitors of *Lycopithecus* 11

§ 1. Climate and landscape in Oligocene and Miocene 11–12

§ 2. Herbivorous animals 12–13

§ 3. Anthropoids—stature of *Lycopithecus* 14

§ 4. His competitors and enemies 14–16

§ 5. Summary 16–17

CHAPTER III

Physical Differentiation of Man 18

§ 1. The erect gait; specialization of legs and feet 18–19

§ 2. —of the hands and arms 19–20

§ 3. Reduction of teeth and jaws; enlargement of cranium . . 20–21

§ 4. Alimentary canal 21

§ 5. Man's relatively naked skin 21–25

§ 6. A dominant animal throughout the world 26

§ 7. *Lycopithecus, Proteranthropus, Homo* 27

§ 8. Modern races of *Homo*; the Nordic sub-race; variation of individuals 27–30

CHAPTER IV

PAGE

Cultural Consequences 31

§ 1. Social life of the pre-human stock formed by co-operation in hunting 31–32
§ 2. Beginnings of speech 32–34
§ 3. Development of intelligence 34–35
§ 4. The making of fire 35–37
§ 5. Cannibalism 37–38
§ 6. Loss of seasonal marriage 38–39
§ 7. Summary of Chaps. III and IV 39–41

CHAPTER V

Mental Differentiation of the *Hominidæ* 42

§ 1. Heredity, Adaptation, Accommodation 42–43
§ 2. Mentality of the original stock 43–45
§ 3. Conditions of our differentiation: the hunting life; geographical diffusion; social life; imaginations concerning the supernatural . 45–49
§ 4. Primal Society: types of mammalian gregariousness. Supposed views of Darwin, Atkinson, etc. 49–52
§ 5. Organization of the hunting-pack persists in all human societies 52–53

CHAPTER VI

Psychology of the Hunting-Pack 54

§ 1. Wolf-pack the nearest parallel 54
§ 2. (1) Interest in the chase and in killing; (2) gregariousness; (3) various modes of sympathy; (4) aggressiveness; (5) claim to territory; (6) recognition of leaders; (7) submission to the pack; (8) emulation; (9) order of precedence; (10) strategy and persistence in pursuit; (11) division of prey 54–61
§ 3. Effects of the hunting life on intelligence 61–63
§ 4. Contrasted character of the herbivorous herd 63–65

CHAPTER VII

Acquisitions in Culture and Custom 66

§ 1. Antiquity of human differentiation; its later forms . . . 66–67
§ 2. Evidence of stone artefacts; full adaptation to the hunting life prior to the Neolithic stage 67–70
§ 3. Slow progress of culture 70–71
§ 4. Constructive ability 71
§ 5. Growth of language 71–73

PAGE

§ 6. Growth of customs; marriage 73–74
§ 7. Property 74–76
§ 8. War 76–77
§ 9. Amusements 78
§ 10. Laughter and lamentation 78–79

CHAPTER VIII

Moralization of the Hunters 80

§ 1. Anthropoid and human passions 80–81
§ 2. Friendliness and benevolence 81–82
§ 3. Moral sense—Darwin's account of it 83–84
§ 4. How influenced by customs 84–86
§ 5. —and by the growth of industry and intelligence . . . 86–88

CHAPTER IX

Influence of the Imaginary Environment 89

§ 1. Growth of imaginative beliefs—primitive Man 89–91
§ 2. How they arose 91–93
§ 3. Evil and good of these beliefs 93–94
§ 4. Influence of wizards and priests on early government . . . 94–97
§ 5. —and on civilization 97–98

Note on the locus of the differentiation of *Lycopithecus* (p. 5) 98

Index 99

THE ORIGIN OF MAN

CHAPTER I

AN HYPOTHESIS CONCERNING OUR ORIGIN

§ 1. MAN WAS DIFFERENTIATED FROM THE ANTHROPOIDS BY BECOMING A HUNTER

THAT the human species, as we now see it, with its several races, Mongolian, Negro, Mediterranean, represents a Family of the Primates is generally agreed; and there is evidence that the Family formerly comprised other species that have become extinct. Our nearest surviving zoological relatives are the gorilla and chimpanzee, the orang and (at a still further remove) the siamang and gibbons; and in spite of the fundamental anatomical resemblance between those apes and ourselves, our differences from them are so great that we cannot wonder at the incredulity with which the doctrine of our consanguinity was first received. Even A. R. Wallace thought that the descent of the *Hominidæ* could not be explained by natural causes; yet we cannot regard our existence as a sort of miracle.

It is the differences between Man and his nearest relatives that have to be accounted for; by derivation from a common stock only his resemblance to them can be understood: heredity explains his nature only in so far as he is an ape. The differences in detail are, indeed, innumerable; but taking the chief of them, and assuming that minor characters are correlated with these, it is the argument of this essay that they may all be traced to the predominating influence of one variation operating amongst the original anthropoid conditions. I do not deny that other causes may have co-operated, but propose to consider how far that one will carry us toward an explanation of the facts, namely, all that we know of the characteristic physical and moral nature of Man. The determining variation was the adoption of a flesh diet and the habits of a hunter in order to obtain it. Without the adoption of a flesh diet there could have been no hunting; but a flesh diet obtained without hunting (supposing it possible) could have done nothing for the evolution of our Family. The adoption of the hunting life, therefore, was the essential change upon which everything else depended. We need not suppose that a whole ancestral species varied in this way; it may have been enough that a few of the common anthropoid stock should do so, provided that the variation was advantageous and was inherited.

Such a change from the frugivorous to the hunting life must have occurred at some time, since Man is everywhere more or less carnivorous, and agriculture is a comparatively recent discovery; the earliest known men were hunters; weapons are amongst the earliest known artefacts. And it is not improbable that the change began at the anthropoid level; because, although extant anthropoids are mainly frugivorous, yet they occasionally eat birds' eggs and young birds; the gorilla is said to eat small mammals, and in confinement they all readily take flesh-food; whilst other Primates (*Cebidæ*, macaques and baboons) eat insects, arachnids, worms, frogs, lizards birds; and the crab-eating macaque (*M. cynomolgus*) collects a large portion of its food upon the Malay littoral. Why, then should not one ape have betaken itself to hunting? Variety o diet, moreover, is not peculiar to the Primates: it is found ir other Orders—marsupials, bats, rodents; whilst amongst car nivora the bears are nearly all omnivorous—the Arctic bea feeding chiefly on seals, porpoises and fish, the grizzly anc the American black bear being extensively carnivorous bu also consuming a good deal of vegetable food, the brown bear in its many varieties adapting its diet to the region in which it lives, and the Indian sloth bear (*Melursus*) confining itself to fruit, insects and honey.

We are not to suppose that our early ancestors became at once exclusively carnivorous: so sudden a change might have put too great a strain on their digestive economy. Even amongst hunting tribes a mixed diet is the rule; and everywhere the women collect and consume fruits and roots. But if at first omnivorous, our ancestor (I conjecture) soon preferred to attack mammals and advanced at a remote date to the killing of the biggest game found in his habitat. Everywhere savage hunters do so now: the little Semang kills the tiger, rhinoceros, elephant and buffalo; and thousands of years ago, in Europe, men slew the reindeer and mammoth, the horse and the bison, the hyæna and the cave-bear. It is true they had weapons and snares, whilst the first hunters had only hands and teeth. These however were formidable weapons of aggression; and their power must have greatly increased if a number of apes cooperated in the chase, forming a hunting-pack, as a sort of wolf-ape (*Lycopithecus*).

In a friendly communication it has been said that the great difficulty of the above hypothesis lies at the beginning of the adventure, in the first change of the feeding habit and the good success of it. I admit this. The gait of a gorilla or chimpanzee upon the ground (the orang is still more arboreal) is an awkward shuffle in which they help themselves along with their long arms; in open forest they move faster, swinging themselves forward by the lower boughs of trees. But neither plan is well adapted to hunting. We cannot, indeed, confidently assume that the anthropoids of the Upper Oligocene (if our differentiation began then) had just the same mode of progression on the ground as those now extant; but these supply the only clue to their habit; and if it was somewhat similar, they were not at such a disadvantage with their contemporaries as they would be if they had to contend with the herbivora and carnivora of our day. For, according to Prof.

Osborne, animals of the Lower Miocene, both herbivora and carnivora, were clumsy and slow-moving[1]. The average pace of the Mammalia, herbivores, carnivores and ourselves, has greatly improved during the last two or three million years: a natural result of competition. Again, what we know of the anthropoid style of fighting suggests that it is a poor preparation for attacking prey. Mr Hornaday says that orangs in captivity are quarrelsome and, when fighting, try (1) to seize and bite an adversary's fingers, (2) attack his face and try to bite his lips[2]. Similarly, the chimpanzee, fighting with a leopard, tries to seize its paws and bite the claws off. If our progenitor naturally fought in this way, he must have adopted some other plan in attacking (say) one of the primitive hornless deer—must have found the throat or spine; but this he may have learnt in capturing smaller prey. It is not improbable that the adventure of hunting for animal food was attempted more than once by Primates and failed, but once, in a happy conjuncture of circumstances, was successful.

The change from a fruit-eating to a hunting life, subserved the great utility of opening fresh supplies of food; and possibly a shortage in the normal supply of the old customary diet was the immediate occasion of the new habit. If our ape lived near the northern limits of the tropical forest and a fall of temperature there took place, such as to reduce (especially in winter) the yield of fruit and other nutritious vegetation on which he had mainly subsisted, famine may have driven him more frequently to attack other animals[3]; whilst more southerly anthropoids, not suffering from the change of climate, continued in their ancient manner of life. In Central Europe, during the Miocene period, the climate altered from subtropical to temperate with corresponding changes in fauna and flora; hence it formerly occurred to me that perhaps the decisive change in the life of our Family occurred there and then. Good judges, however, put the probable date of the great differentia-

[1] *The Age of Mammals*, p. 249.
[2] *Mind and Manners of Wild Animals*, p. 272.
[3] Suggested to me by Mr G. A. Garfitt.

tion much earlier, in the Oligocene[1]. Indeed the occurrence of a chimpanzee (*Dryopithecus*) in a Miocene formation of Europe may be held to indicate that the anthropoid stock had already broken up. But in the Oligocene I cannot find that any extensive change of climate has been detected. As, however, not much is known of the condition of Central Asia at that time, it is possible that a considerable elevation of land took place there. The Himalayas, indeed, attained their present elevation only in the Pliocene; but the area had been rising for a very long time; and if it reached in the Oligocene the height of only five or six thousand feet, that may have sufficed to reduce in the area affected the supply of the customary anthropoid food so far as to make hunting a profitable or necessary alternative. [See Note at p. 98.]

Awaiting adequate evidence for such conjectures, there remains, in the last resort, "spontaneous" variation: that is

[1] Estimated duration of the Cainozoic Period, assuming that the thickness of the deposits is about 63,000 feet, and that deposits accumulate at the rate of 1 foot in 100 years. Drawn to the scale of 1 mm. to 100,000 years. The estimate is given and explained by Prof. Sollas in the *Quarterly Journal of the Geological Society* (1909), LXV. The "tree" is based on that given by Sir A. Keith in *The Antiquity of Man*, p. 509.

If we suppose the differentiation of the *Hominidæ* to have begun before the close of the Oligocene, about (say) 3,500,000 years are allowed for the evolution of the existing species of Man. All these reckonings are provisional.

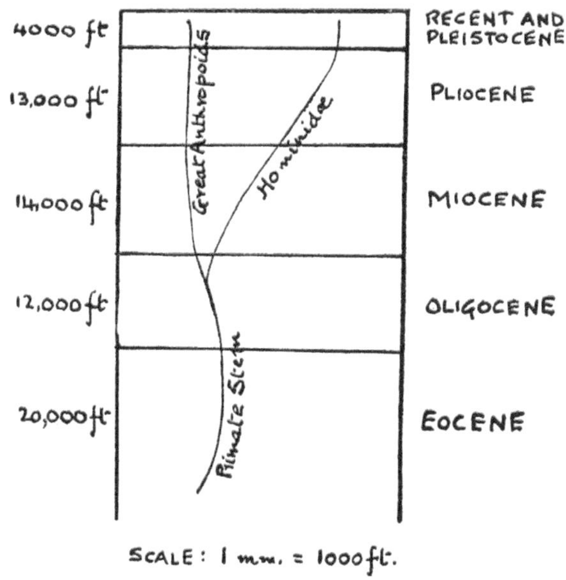

to say, from causes which are at present beyond our knowledge, the fateful ape did in fact prefer animal food so decisively as to begin the hunting for it. That being granted, the rest of the history was inevitable. The new pursuit was of a nature to engross the animal's whole attention and coordinate all his faculties; and to maintain and reinforce it, his structure in body and mind may reasonably be supposed to have undergone rapid modification by natural selection; because those individuals that were in any organ or faculty adapted to the new life had an advantage which might be inherited and gradually increased[1].

[1] That Man was from the first a hunter has been suggested by several authors; but the consequences of the assumption have never (as far as I know) been worked out. A. R. Wallace, in *Darwinism*, p. 459, has the following passage: "The anthropoid apes, as well as most of the monkey tribe, are essentially arboreal in their structure, whereas the great distinctive character of man is his special adaptation to terrestrial locomotion. We can hardly suppose, therefore, that he originated in a forest region, where fruits to be obtained by climbing are the chief vegetable food. It is more probable that he began his existence on the open plains on high plateaux of the temperate or sub-tropical zone, where the seeds of indigenous cereals, numerous herbivora, rodents, game-birds, with fishes and molluscs in the lakes and rivers and seas supplied him with an abundance of varied food. In such a region he would develop skill as a hunter, trapper or fisherman, and later as a herdsman and cultivator—a succession of which we find indications in the palæolithic and neolithic races of Europe."

Prof. MacBride, in his popular introduction to *Zoology*, p. 84, also traces the specialisation of Man to the hunting life.

My friend Mr Thomas Whittaker has sent me the following extract from Comte's *Politique Positive*, I, pp. 604–5: "L'obligation de se nourrir d'une proie qu'il faut atteindre et vaincre, perfectionne à la fois tous les attributs animaux, tant intérieurs qu'extérieurs. Son influence envers les sens et les muscles est trop évidente pour exiger ici aucun examen. Par sa réaction habituelle sur les plus hautes fonctions du cerveau, elle développe également l'intelligence et l'activité, dont le premier essor lui est toujours dû, même chez notre espèce. A tous ces titres, cette nécessité modifie aussi les races qui en sont victimes, d'après les efforts moins énergiques, mais plus continus, qu'elle y provoque pour leur défense. Dans les deux cas, et surtout quant à l'attaque, elle détermine même les premières habitudes de co-opération active, au moins temporaire. Bornées à la simple famille chez les espèces insociables, ces ligues peuvent ailleurs embrasser quelquefois de nombreuses troupes. Ainsi commencent, parmi les animaux, des impulsions et des aptitudes qui ne pouvaient se développer que d'après la continuité propre à la race la plus sociable et la plus intelligente. Enfin, la condition carnassière doit aussi être appreciée dans sa réaction organique. Une plus forte excitation, une digestion moins laborieuse et plus rapide, une assimilation plus complète produisant un sang plus stimulant: telles sont ses propriétés physiologiques. Toutes concourent à développer

§ 2. Natural Selection

Having thus appealed to the principle of natural selection as controlling the evolution of Man, I must explain what is to be understood by it. In the first place, it has nothing to do with the causes of variation. Much interesting and instructive work has been done by Biologists upon the structure of cytoplasm and the possible results of combination and recombination among its constituents, chromosomes and genes, and upon the conditions which increase or decrease variation in resulting generations. But that in some way variations occur is here assumed, and we are concerned only with what happens to them afterwards. Nor do the Mendelian laws of inheritance affect this problem; for in whatever way an animal is constituted by inheritance, having been born it must either live or die; and it is with this alternative that natural selection is concerned. If the animal is not sufficiently adapted to the conditions of life, interuterine, natal and environmental, climatic or biological, to live at least until the age of propagation, it must die without offspring: it is eliminated.

But it has been urged that the condition of such elimination is not well expressed by the phrase "survival of the fittest." Not only the fittest but many less fit can, and do, normally survive; for that they need only reach a certain standard of fitness. So much is plainly true. What shall be the standard of the least unfit, however, must depend upon the severity of the conditions of life, competition for food and mates, self-maintenance against enemies, rivals, disease and whatever else may be inimical to their welfare. After such a change of life as I have supposed on the part of our ancestral ape, the struggle probably was very severe, and the standard of fitness was very high.

Further, it has been urged that many characters that seem to us very important in the classification of animals, or in

les fonctions supérieures, soit en augmentant l'énergie de leurs organes, soit en procurant plus de temps pour leur exercice."

For the views of Mr Ch. Morris in *Man and His Ancestors* (New York, 1900) see the Preface to this edition.

the determination of human races, cannot be shown to have any survival value, and therefore cannot be explained by natural selection; and this also seems to be true. What adaptation is involved in the distinction between long-heads and round-heads, upon which Anthropologists have done so much careful statistical work? What survival value can be assigned to the Negro's heel, or to the Kalmuck's slanting eye, or to the remarkable differences in the hair-section of our conspicuous races? But it is not with such characters that we have to do in explaining the adaptation of Man to the life of a hunter. Consider the erect gait, the modification of legs and feet, of arms and hands, social life, language, intelligence, the discipline of the pack. These are the things that I attribute to natural selection; and will anyone deny that they are adaptations to the hunting-life of Man and conditions of all his development?

§ 3. An hypothesis is an inference from the facts it is presented to explain. Hypothetical reasoning is almost universal in science and very common in every-day life; yet it is often regarded with a dull suspicion that can only result from misunderstanding. The form of such reasoning seems to be deductive; the hypothesis is stated, and the facts seem to be inferred from it; and our frequent resort to this mode of stating a case led Whewell to remark that "Man is prone to become a deductive thinker." But the truth is that the argument is inductive: the form of statement turns the psychology of it upside down; for the argument really is that the hypothesis may be inferred from the facts. What usually (perhaps always) happens, I believe, is that one or a few facts may suggest a common cause, or schema, as their explanation; then this explanation is constituted an hypothesis, and one goes on to show how, if true, it will lead to all the given facts and to as many others as possible within the sphere of investigation. But in saying that a few facts suggest a common cause, we mean that this cause may be inferred from them; and, extending it to more and more facts, we mean that it may be inferred from them also. The facts from which this hunting hypothesis were first inferred by me were the modification of our legs and arms from those of the simian type, the upright gait, intelligence, social organization (like that of wolves) and freedom of movement beyond the tropical forest[1]. These changes are such as might be expected to follow if an anthropoid ape should have become a hunter. Then, assuming that one did so, certain other changes (as below) may be deduced; that is to say, from these further changes also the original hunting life may be inferred.

Since each of these inferences is from an effect to a possible cause, none of them is conclusive. When an effect is conceived in a general way, it often happens that it may be explained by more than one cause. But each inference raises some probability in favour of the cause, and as one instance is added to another the probability increases; and at the same time the probability that any other cause would explain all the facts equally well grows less and less. As we cannot attach any numerical values to the probabilities severally, we cannot exactly estimate their value altogether. Each reader must make his own estimate as best he can. For my part I think the total probability may fairly be put at more than a half.

It is a great advantage in verifying an hypothesis when other hypotheses to the same purpose have been advanced, and it is possible to refute them; for it may then appear that not only is the hypothesis in some degree probable, but that, as the alternatives go down one by one, it is probably the only valid one. But in the present case no such help is offered; for (as far as I know) there is no other hypothesis (limited to the natural order) that attempts to explain how the human race came to exist.

To refute the argument one may show (1) that the hypothesis cannot be inferred with any probability from this, that and the other stated fact; (2) that there are other differences between ourselves and the anthropoids (of equal weight with those I mention) from which the hypothesis cannot be in-

[1] *Metaphysics of Nature*, ch. xv, § 3.

ferred; (3) that some change other than the adoption of the hunting life, would, in the conditions of some anthropoid's life, explain all the facts equally well or better. I invite attention to these considerations.

If the problem of the causes of our differentiation is to be dealt with at all, there is no other method at our command except such an accumulation of probable inferences from the known facts of our present condition in comparison with that of the apes. The subject is not open to observation or experiment. It has been said that the true method is to compare all that we know of primitive Man, fossil Man and so forth. I have taken account of these things so far as they throw any light upon the inquiry; but consider how little we know of fossil Man and his congeners. Suppose we found in the later Miocene a complete skeleton of a Primate with human-like characters: it would be a new species; no one could be confident that it stood in the line of our ancestry. Suppose we should find a complete series of skeletons, one for every 200,000 years from the end of the Pliocene back to the Oligocene, and that experts should agree that they represented the "orthogenic" evolution of *Homo sapiens*: we should hardly be any nearer a solution of our present problem. For the remains would not show the conditions under which the differentiation began and was maintained, but would merely add to the data upon which an hypothesis might be constructed. In short nothing can be done in the matter except by thinking, by trying to think what is most probably indicated by all the facts within our knowledge. The leaders of scientific investigation do not shrink (I observe) from thinking courageously or even audaciously. But a good many people, relying too much on their own experience, adopt the sentiment of that mighty verse:

Thinking is but an idle waste of thought.

CHAPTER II

PREY AND COMPETITORS OF *LYCOPITHECUS*

§ 1. If we suppose the differentiation of the *Hominidæ* from the *Anthropoidea* to have begun in the Upper Oligocene, and that the decisive change was initiated by some ape that adopted the life of a hunter, it is interesting to consider what the world was like in which he lived, what sort of animals surrounded him, what animals probably became his prey, and what were his rivals in the chase[1].

The surface of the planet was less mountainous than at present; in Europe the Pyrenees had risen, but the Alps were only beginning to rise; and in Asia the Himalayas began to dominate the world only in the middle of the next epoch, the Miocene. The distribution of land and water, too, was very different in the Oligocene from that which we now see: Europe was divided from Asia by a broad gulf stretching from the Indian Ocean to the Arctic Circle, and an arm of this gulf toward the west submerged a great part of Central Europe; Asia was broadly connected with North America, where now the sea penetrates between Siberia and Alaska; Africa had no connection with either Europe or Asia; North and South America were separated—perhaps at Panama. In the Miocene, Europe, Asia and Africa became united. These physiographic changes may have affected climate; for during the Eocene tropical conditions prevailed far to the north, and coal-beds were laid down in Alaska; but from the Oligocene onwards there was a gradual fall of temperature, slow at first, but ending (for the present) in the cataclysms of the Glacial Period. There was also a decrease in some regions of atmospheric moisture, which determines the density of vegetation.

[1] The contents of this chapter lie outside my own studies, and have been taken from various books of Geology and Palæontology: I must especially mention Prof. Osborne's *Age of Mammals* (1910) and Prof. Scott's *Land Mammals in the Western Hemisphere* (1913). I have also profited by inspecting the Palæontological Gallery at South Kensington with the help of its excellent Guide-Book.

In its general character the vegetation was similar to that which now prevails in tropical, subtropical and temperate regions of the world. The species of plants now existing had not yet arrived; but of the same genera and Families as those we see, conifers, palms and dicotyledonous flowering plants crowded the forests and overhung the rivers. The forests were more extensive and continuous than ours outside the tropics; for by degrees browsing animals, feeding down the young trees, check the renovation of forests and clear open spaces, where grasses grow; elevations of the land and changes of temperature limit the northern or southern extension of certain kinds of plants, and a failure of humidity starves all the larger kinds; converting, at successive stages, forest into steppe and steppe into desert.

§ 2. Animals, especially mammalia, with which chiefly we are concerned, were, at the close of the Oligocene, very different from any that now roam the lands; all the species, most genera, many Families and some whole Orders have since disappeared. But there were plenty of animals to eat and a good many to dread. Until we know the neighbourhood in which our ape's adventures began, nothing precise can be said of his circumstances. Probably it was somewhere in the Old World, and probably it was in Asia. Unfortunately, we know nothing of the zoological antiquities of Asia until the early Miocene, and even then a very small selection of what must have existed, because geologists have hitherto explored a very small part of the continent—a few beds in north-western and northern India and in Burmah and Mongolia. But there is so much evidence of the migrations of animals in successive ages of the Tertiary Period, that any remains from the Oligocene and Miocene will help us to understand what sort of neighbours our remote ancestors had to live amongst.

For prey there was great variety of birds and reptiles (everywhere eaten by savages) and fishes; but we confine our account to the mammalia, which he may be supposed to have pursued afoot. Of these the most important are the hoofed animals, which fall into two great groups, perhaps not closely connected—the odd-toed (Perissodactyls) and the even-toed (Artiodactyls). During the Oligocene there lived in Europe, or in North America, or in both—and, therefore, probably in Asia—numbers of the odd-toed group: tapirs; rhinoceroses of several species, some without horns, some with, some amphibious (Amynodonts), all smaller than their modern representatives; chalicotheres, strange beasts something like horses, but having, instead of hoofs, claws on their toes—perhaps they survived in China into the Pleistocene; small predecessors of the horse with three toes on each foot; titanotheres, hugest animals of their age, extinct in the middle of it—something like the rhinoceros and nearly as big as an elephant (*Brontotherium*). Of the even-toed group, pig-like animals abounded, and some true pigs appeared; entelodonts, or giant-pigs, were common; anthracotheres, somewhat pig-like in size and shape; ancestral camels about the size of sheep were to be had in North America; oreodonts, unfinished-looking creatures of many species; primitive deer and other ruminants, small in size and not having yet grown any horns. In Europe, during the Upper Oligocene, cœnotheres, small and graceful animals, lived in large herds around the lakes. There were also primitive proboscidia about half the size of modern elephants; many insectivores; and, amongst rodents, beavers and tailless hares. Generally, animals of this age that have left descendants were smaller than their modern representatives; and notably their brains were smaller.

In the Lower and Middle Miocene there appeared also horned cervuline deer, chevrotaines, and horned antelopes; dinotheres and mastodons, probably from Africa; primitive hedgehogs, moles and shrews; and in the Upper Miocene, hipparion, true hares, several varieties of hornless giraffe, true deer, and ancestral sheep. True horses and cattle are first known from Pliocene beds; but it is needless to follow the story further: the fauna becomes more and more modern in its character, and uncouth forms die out.

§ 3. Anthropoids are first met with in the Miocene and in Europe: pliopithecus, allied to the gibbons, in the Lower; and dryopithecus, related to the chimpanzee, in the Middle Miocene; but they are believed to have come from Asia. There, in Pliocene beds of the Siwaliks (southern foot-hills of the Himalayas), occur the orang and chimpanzee, besides macaques, langurs and baboons. Since the orang is now found only in Borneo and Sumatra, and the chimpanzee only in Africa, southern or central Asia seems to have been the country from which the anthropoids dispersed; and this seems to be the chief positive ground for believing that the human stock began to be differentiated in that region—perhaps from a variety of anthropoid living on the northern side of the common habitat. Since, again, by the Middle Miocene a chimpanzee form had already migrated into Europe, it may be assumed that the orang was already distinct from it (and perhaps had spread eastward): the differentiation of these genera must, therefore, have happened earlier; and, therefore, also the differentiation of the human stock; so that this event cannot be put later than some time in the Oligocene.

How big was Lycopithecus to begin with? The answer to this question must affect our view of his relations both to prey and to enemies. Inasmuch as the three extant anthropoids and Man are all of about the same size, there is a presumption that their common ancestor was in stature superior to the gibbons and to the largest monkeys—in fact, a "giant ape" (to borrow a term from Dr Keith). Dryopithecus "was smaller than the chimpanzee, but much larger than the gibbon[1]." Awaiting further evidence of fossils, which is much to be desired, it is probable, on the whole, that Lycopithecus weighed less on the average than modern man, but more than the wolf.

§ 4. As to competitors and aggressive enemies, there were snakes and crocodiles; but, confining our attention to carnivorous mammals, the time seems to have been favourable

[1] A. Keith, *The Human Body*, p. 58.

to the enterprise of a new hunter. By the middle of the Oligocene, the ancient Creodonts (primitive flesh-eaters which had flourished in the Eocene) were nearly extinct, represented in the deposits by their last surviving Family, the Hyænodonts. Ancestors of the modern carnivores, such as may be called by anticipation dogs and cats, derived (according to Prof. Scott) from the Creodont Family of the *Miacidæ*, were becoming numerous, but for the most part were still of small size. Apparently, the primitive dogs and their allies must, for some time, have been more formidable adversaries than the primitive cats, especially if we suppose them to have already begun to hunt in pack; and this is not improbable, both on account of their structure and because several distinct species and even genera, now extant, have that habit—such as wolf, jackal, dingo, dhole, Cape hunting-dog, etc. In the Upper Oligocene of North America, occurs a dog as big as a large modern wolf, and in Europe the bear-like dog, Amphicyon, of about the same size, but said to have been clumsy and slow-moving. There were several other dog-like species; they continue in the Miocene, and some of them increase in bulk; but true modern dogs or wolves (*Canis*) do not appear before the Pliocene. Then, too, first occur true bears (*Ursus*); hyænas in the Upper Miocene. "Cats" belong to two sub-Families: (i) the true felines, our modern species and their ancestors; and (ii) the machærodonts, or sabre-toothed cats. The latter first appear in North America in the Lower Oligocene; the former in Europe in the Middle Oligocene. The sabre-toothed are so called from their thin, curved upper canines; which were so long (3 to 6 inches) that it is not easy to understand how they could open their mouths wide enough to bite with them. That they were effective in some way (perhaps by laceration) is proved by the fact that machærodonts, first appearing in the Lower Oligocene, increased in numbers and diversity of species for ages, and some of them in bulk. In North America, in the Upper Oligocene, one species was as large as a jaguar, and some of the biggest and most terrifying were contemporary with Man, and only became extinct in the Pleistocene. Their limbs were relatively shorter and thicker than those of the *Felinæ.*

These, the true cats, at first progressed more slowly than the *Machærodontidæ*; but in the Siwalik deposits (Pliocene) there occur, along with machærodonts, forms resembling the leopard and the lynx, with others as large as tigers. The largest of all this group seems to have been the cave-lion, perhaps a large variety of the common African lion, which also lived with Man in Europe in the Pleistocene. These were serious competitors in the hunting-life of Lycopithecus and of primitive Man; and the effect of such competition in exterminating inferior forms is shown by the fate of the carnivorous marsupials of South America (allied to *Thylacinus*), which were the predatory fauna of that region, until in the Pliocene, North and South America having become united by continuous land, cats and dogs came in from the northern continent and put an end to them; and also by the fate of the creodonts, which in the Oligocene seem everywhere to have been exterminated by the new carnivores. In both cases the beaten competitors were very inferior in the size and complexity of their brains; and if Man has succeeded in the struggle for life against the same foes, in spite of his inferior bodily adaptation, it is probably due to his very superior brains. This may also be the reason why modern Man (*Homo sapiens!*), wandering everywhere over the world, has everywhere exterminated such experiments in human nature as Pithecanthropus, Eoanthropus, and Neanderthalensis; as others are soon to follow them into the Hades of extinct species.

§ 5. These few pages give a ridiculously faint sketch of the animal world amidst which our remote ancestors began their career. But it may serve to indicate that there was always plenty to eat if you could kill it, and plenty of rivals who wanted their share. After the disappearance of the dinosaurs at the close of the Cretaceous period, the mammalia, already numerous, developed rapidly, and spread in ever multiplying numbers and diverging shapes over the whole area of the land.

We may take it that from the Middle Eocene (at least) onwards the earth has always been as full of wild beasts as it would hold. To understand what it was like in the Middle Oligocene, one should read the adventures of hunters in South Africa seventy or eighty years ago (their verisimilitude is vouched for by Livingstone[1]), before a gun in the hands of every Kaffir had begun to thin the vast herds that then covered the whole landscape, and in whose numbers the wild hunters and the lions could make no appreciable diminution. The little Boschmans regarded themselves and the lions as joint owners and masters of all the game. The masters fought one another, indeed; but there was no necessity to fight, for there was more than enough for both: lions were then sometimes met in gangs of ten or a dozen. Game throughout the Cainozoic ages was abundant and of all sizes: many small, many middle-sized and some prodigious. Even in the Eocene, some of the Amblypoda (*Dinoceras*, Am.) and of the Barypoda (*Arsinotherium*, Af.) were as big as rhinoceroses; in the Oligocene, Titanotheres not much smaller than elephants; in South America, in Miocene and Pliocene times, the Toxodonts; in the Pleistocene, Ground-sloths of huge bulk, and Glyptodonts. Of Families still represented amongst living animals, dinotheres and mastodons occur in the Miocene; and elephant, rhinoceros, hippopotamus, giraffe have abounded from the Pliocene to recent times, in many species, over most of Africa and the northern hemisphere. Even the marsupials in Australia produced a species (*Diprotodon*) as large as a rhinoceros with a skull three feet long. Any one of these would have been a week's food for a whole pack of hunters, if they could kill it—as we may be sure they could.

[1] *Travels and Researches in Western Africa*, ch. vii.

CHAPTER III

PHYSICAL DIFFERENTIATION OF MAN

Somewhere, then, perhaps in central Asia, and perhaps in the latter part of the Oligocene Period, I suppose that an anthropoid, originally for the most part frugivorous, became a hunter for animal food. And if that really happened he would probably, in adaptation to the new mode of life, have undergone in the course of ages many changes in his other habits and in the structure of his frame, approaching more and more to the human type.

§ 1. Darwin says: "As soon as some ancient member of the great series of the Primates came to be less arboreal, owing to a change in his manner of procuring subsistence, or to some change in the surrounding conditions, its habitual manner of progression would have been modified: and thus it would have been rendered more strictly quadripedal or bipedal. Baboons frequent hilly and rocky districts, and only from necessity climb high trees; and they have acquired almost the gait of a dog. Man alone has become a biped[1]." Not only the erect attitude, which is widely attained amongst the arboreal Primates[2], but the erect gait must of course be understood as the normal mode of progression. The gibbon is the only other Primate that attempts this, and only for short distances and with much careful balancing of himself with his arms: being in fact more truly arboreal than the chimpanzee. The erect gait, says Darwin, would naturally result "from some change in the manner of procuring subsistence, or some change in the surrounding conditions"; and such a change in the manner of procuring subsistence, would have been the adoption of the hunting life and the practice of following prey afoot upon the ground.

[1] *Descent of Man*, 2nd ed., p. 51.   [2] F. Wood Jones, *Arboreal Man*.

It may, indeed, be supposed that at first prey was sometimes attacked by leaping down upon it from the branch of a tree, as leopards often do; but such simple strategy could never have made our stock the dominant one of the world: nothing could do this but the gradual attainment of the erect gait adapted to running down our prey. The less our ancestor trusted to trees the better for him if he was to fulfil his destiny. It is not, indeed, in dense forest that game most abounds. Where some elevation of the ground had dried the soil and thinned the trees, where herbivores had browsed down and trimmed the borders of the forest, or along the shores of shrinking lakes, would lie the most favourable hunting-grounds. And to form and maintain the erect gait the numerous modifications of structure necessary to it, whenever from time to time they occurred, were preserved and accumulated by natural selection: namely, the curving of the vertebral column, the balancing of the head on a relatively slender neck (still imperfect in Neanderthalensis), the broadening of the pelvis, changes in joints, bones and muscles of the legs, the lengthening of the leg, and specialization of the feet—in which the heel is developed more than in the gorilla, the sole flattened and the great toe lengthened and laid parallel with the others.

§ 2. The specialization of the legs and feet as it proceeded made possible the specialization of the hands: being gradually rid of the task of assisting locomotion, whether in trees or on the ground, they were used in grappling with prey, seconded by massive jaws and powerful canine teeth. After a time rude cudgels and stones were brought to the encounter, and in the course of ages such means of offence began to be altered into weapons such as stone axes and spears, that might be called artefacts. Becoming more and more erect and bipedal, "they would have been better able to defend themselves with stones or clubs, to attack their prey or otherwise to obtain food," says Darwin[1]; and I must point out that he seems here to assume that Man may have become a hunter at a very

[1] *Descent of Man*, 2nd ed., p. 52.

early date. These simple beginnings probably occupied an immense time, perhaps more than half of the total period down to the present. The utility and consequent selection of hands had been great throughout; but their final development may be referred to the making and using of weapons fashioned according to a mental pattern. Those who had the best hands were selected because they made the best weapons and used them best; but we know from the remains of several palæolithic stages of the art of manufacturing implements how very slowly the art improved.

Along with specialization of the hands went a reduction in the length and massiveness of the arms with such modifications as rendered the hands easier to control and more adroitly serviceable. Comparatively weaker arms may have been disadvantageous in directly grappling with prey; but it was necessary to the runner in order to lessen the weight and cumbersomeness of the upper part of the body and to improve his balance and agility. The change may also have been beneficial by affording physiological compensation for the lengthening and strengthening of the legs. And as soon as unwrought stones and clubs came into use there was mechanical compensation for the shortening of the arms. The result is an adaptive co-ordination of the total structure to the life of a two-footed hunter. Why is Man a running animal? Is it for the advantage of running away? To run away is sometimes useful, but it is not characteristic of Man: rather to run to the attack and to pursue. Accordingly, though fairly swift, he is not amongst the swiftest animals; but he is very long-winded and indefatigable, and in that, as in many other things, he resembles the dogs and wolves.

§ 3. Darwin says: "The early male forefathers of Man were probably furnished with great canine teeth; but as they gradually acquired the habit of using stones, clubs, or other weapons for fighting with their enemies or rivals, they would use their jaws and teeth less and less. In this case the jaws, together with the teeth would become reduced in size, as we may feel almost sure from numerous analogous cases[1]." Eoanthropus is generally believed to have retained the ancestral canines.

With the reduction of the jaws, the profile began to approach the orthognathous type; and it progressed further in that direction on account of accompanying changes in the skull. The skull became less thick and rough, (*a*) because, as the hands (using weapons) superseded the teeth in fighting, jaws and neck grew less massive, and their muscles no longer needed such solid attachments; (*b*) because the head was less liable to injury when no longer used as the chief organ in combat. At the same time the skull slowly increased in capacity and became vaulted to make room for the brains of an animal which (as we shall see) acquired much knowledge (parietal association area) and lived by the application of its knowledge to the co-ordination of increasingly complex and continuous activities (anterior association area).

§ 4. The extensive adoption by Man of a flesh diet many hundreds of thousands of years ago might be expected to have shortened his alimentary canal in comparison with that of the anthropoids; but not much evidence of it is obtainable. Topinard, giving a proportionate estimate, says that in Man it is about six times the length of the body, in the gibbon about eight times. Sir Arthur Keith, in a private communication with which he has favoured me, says that the adult chimpanzee's intestine is slightly longer than the adult man's, but that the measurements are, for certain reasons, unsatisfactory, and that there have not been enough measurements of adult chimpanzees. We must remember that, on the one hand, the chimpanzee is not exclusively frugivorous and that, on the other hand, it is not likely that Man has been at any time exclusively carnivorous; though the return of large populations to a vegetarian diet by means of agriculture is recent.

§ 5. There is one characteristic difference of Man from the

[1] *Op. cit.* p. 53.

anthropoids which his hunting habits do not clearly explain —his relatively naked skin. Darwin attributed this condition to sexual selection[1]. He argued that, on the one hand, so far as Man has had the power of choice, women have been chosen for their beauty; and that, on the other hand, women have had more power of selection, even in the savage state, than is usually supposed, and "would generally choose not merely the handsomest men, according to their standard of taste, but those who were at the same time best able to defend and support them." Hence, if a partial loss of hair was esteemed ornamental by our ape-like progenitors, sexual selection, operating age after age, might result in relative nakedness. "The faces of several species of monkey and large surfaces at the posterior end of the body have been denuded of hair; and this we may safely attribute to sexual selection." The beard of the male, and the great length of the hair of the head in some races, especially seem due to this cause. The greater hairiness of Europeans, compared with other races, may be a case of reversion to remote ancestral conditions. For as all races are nearly naked, the common character was probably acquired before the several races had diverged from the common stock.

The loss of hair on various parts of their bodies by certain species of monkey, and the beard of males (together with the longer head-hair of women) of our own race are cases that strongly support the ascription of such secondary sexual characters to sexual selection. Yet, going back to the time before the division of modern Man into races (say, 600,000 years), it seems incredible that any women then went unmarried, hair or no hair, if they were healthy (and the unhealthy soon ceased to exist); or that any man went unmarried, if he could do his share in the hunting-field (and, if not, he also soon ceased to exist). No facts observed amongst extant savages—the choice exerted by women, or the polygamy of chiefs—throw much light upon that ancient state of affairs. There were then no chiefs: the hunt-leader of pack or clan

[1] *Descent of Man*, 2nd ed., pp. 595–604.

had no authority but his personal prowess, no tradition of ancestry or religion, nor probably the prestige of magic, to give him command of women. Unless, at that time, relative nakedness was strongly correlated with personal prowess in the male and efficiency in the female, it is difficult to understand how it can have been preserved and increased by sexual selection. Forgive me for adding an unkind remark: if the selection of women for their beauty has gone on for hundreds of thousands of years, and has had a cumulative effect upon the race, is not the result disappointing? Go into the street and look. That "women have become more beautiful, according to the general opinion, than men," is not an objective, truly æsthetic judgment, but one determined by causes of which "general opinion" is falsely unconscious. Schopenhauer[1] thought that men are better looking than women; and of average specimens this seems to be true; though, to be sure, he was a sort of misogynist.

Another explanation of Man's nakedness was suggested by Thomas Belt, based on the parallel case of certain races of naked dogs, namely, that he is the better able to free himself from parasites[2]. Darwin mentions this hypothesis and, in a footnote, cites in its favour "a practice with the Australians, when the vermin get troublesome, to singe themselves"; but he says, in the text, "whether this evil is of sufficient magnitude to have led to the denudation of the body through natural selection, may be doubted, since none of the many quadrupeds inhabiting the tropics have, as far as I know, acquired any specialised means of relief[3]." It appears, too, that against the probability of such a result must be set the actual disadvantage of nakedness, as insisted upon by Wallace, who says that savages feel the want of protection and try to cover their backs and shoulders[4]. Still, the disadvantage implied in occasionally feeling the want of protection could not prevent the loss of hair, if this would deliver the race from

[1] *Parerga und Paralipomena*, II, Kap. xxvii.

[2] *A Naturalist in Nicaragua*, ch. xi.

[3] *Op. cit.* p. 57.

[4] *Natural Selection*, pp. 195–7.

serious dangers from vermin; and the force of the argument from the condition of other tropical quadrupeds depends, at least in some measure, upon whether or not there is something peculiar in the case of naked dogs and men.

Belt argues that the naked dogs with dark, shining skins, found in Central America and also in Peru[1], and which were found there at the Spanish conquest, have probably acquired their peculiar condition by natural selection, because they are despised by the natives, and no care is taken of their breeding, and yet they do not interbreed with the common hairy varieties, as usually happens with artificial stocks. The advantage of a naked skin being the greater freedom it gives from ticks, lice and other vermin, the advantage is especially great for a domestic animal living in the huts of savages, where, because they are inhabited year after year, vermin are extraordinarily abundant. The naked dog, then, differs from tropical quadrupeds which are adapted from a dateless antiquity to such vermin as infest them, by having been thrown by human companionship amongst not only strange vermin, but vermin in extraordinarily dense aggregation. Belt would have guarded a weak point in his case, had he explained why naked races of dogs are so scarce. Hairy races may have been more recently domesticated, or bred for their hairiness, or less addicted to an indoor life.

The case of our own forefather also differs somewhat from that of other tropical mammalia; because, by hypothesis, he underwent pretty rapidly such an extraordinary change of life; which may have brought him into circumstances where vermin, formerly negligible, became highly injurious. "Monkeys," as Belt observes, "change their sleeping-places almost daily"; the Orang is said to construct a fresh nest every night; this is also reported of the Gorilla. Not improbably, then, daily change of locality was the practice of the original anthropoid stock, whence we also are descended: thereby avoiding the accumulation of vermin. Did the hunting

[1] Naked races of dogs have also been reported to exist in China, Manila and South Africa; but I can learn no particulars of them.

life introduce a new habit? In the old frugivorous forest life, the custom was to get up into some tree for the night, and within a short radius there were hundreds of trees equally suitable; and, therefore, there was nothing to check the natural preference for a fresh one. When, however, the hunters began to make lairs in open country, there was no such wide choice amongst caves, rock-shelters, or thickets: one might be better than any other for miles around. If, then, several of them settled down there as in a common lair, the circumstances were, for the time, favourable to the multiplication of vermin, and therefore to nakedness of skin, in order the more easily to be rid of them. So perhaps this difference of Man from the anthropoids may be referred to one common cause with all the others—the hunting life. In that life, too, the defilement of blood made fur inconvenient to animals not apt to cleanse themselves, like those in the true carnivorous heredity and tradition.

When we consider how injurious some insects are to vertebrate life, being suspected of having caused in some cases the extinction of species, can it be said that facility in ridding oneself of such vermin as lice and ticks is an inadequate cause of human nakedness, or not one that might outweigh the drawbacks of cold and wet? It is not, however, a cause incompatible with the action of sexual selection, tending to the same result; nor, again, with the preferential destruction of hairy children if ever infanticide was practised. A further possible ground of deliberate selection may have been the mere ambition of differing from other animals; for a tribe on the Upper Amazons is reported to depilate to distinguish themselves from the monkeys, and the wish to be superior to other animals led a tribe in Queensland to pretend that they, unlike kangaroos, etc., have no fathers according to the flesh[1]. Admitting that this last motive can hardly have been primitive, still, our nakedness may be a resultant of several causes: one of them, and perhaps the most important, having been the insanitary consequences of hairiness to Lycopithecus.

[1] W. E. Ling Roth, *North Queensland Ethnology*, Bulletin v, § 81.

§ 6. That the earliest men of whose condition we have any knowledge were hunters supports our hypothesis. Any other view of Man's origin must explain how and when he became a hunter. That all races of men are hunters certainly indicates that the habit had been adopted before the division of races (say) in the Pliocene. But there seems to be no reason to put the great change of habit anywhere nearer than the beginning of our differentiation. The further we put it back the better it explains other modifications.

One of the most obvious consequences of relying upon animal food and hunting for it was that Man became a dominant animal throughout the world. The anthropoids are never found outside of the tropical forests of Africa and Malaya (including Borneo and Sumatra). They feed chiefly on the fruits and other highly nutritious vegetables that, all the year round, are only there obtainable. Although often coming to the ground, especially the chimpanzee and gorilla, they are adapted to living in the trees: that is their home. In contrast with their habits, Man is at home upon the ground, with unlimited range over the whole planet from beyond the Arctic Circle to Tasmania and Tierra del Fuego; because on the ground (chiefly) he everywhere, whilst in the savage stage of culture, finds his food in the other animals whom he hunts and slays. This, then, was the condition of his emancipation from the tropical forest. It is, indeed, conceivable that a frugivorous animal, originally of the forest, should obtain a wider range by taking to a coarser diet of roots and herbage, such as suffices the Ungulates, browsing, or grazing, or digging with their snouts; but this would not have led to the upright gait, or to the big brain, or to any of the marks that distinguish Man. Not advance but retrogression must have followed such a change. Some hunters, however, as they passed into countries where the more nutritious vegetable diet became scarcer and scarcer, may have supplemented their animal food with roots and leaves and other coarse fare; and this may explain the dentition of Neanderthal Man, which is said to be adapted to such food.

§ 7. In the course of development our ancestor must have passed through certain stages of which at present there is no fossil evidence, during which he was at first for ages, say to the Middle Miocene, recognizably an ape (Lycopithecus); then another stage, slowly supervening (and, of course, nowhere definitely distinct) extending perhaps to the Middle Pliocene, during which he approached the human form and faculties and might be called Proteranthropus—*Pithec-anthropus erectus* may be a belated specimen of this transition; and, finally, the stage at which, divided into several species, he is in all of them recognizably *Homo*, and in one of them *Homo sapiens*. At the earlier stages many branches of the tree may have been lost.

Mr Morris thinks that the earliest stage of *Homo* is represented by the Pygmies[1]; but perhaps it may be better to consider these peoples as regressives. He also thinks that a great advance in culture was made during the glacial period[2]: an opinion in which he has been followed by Max Nordau and by Dr McCabe. But it seems to me not very reasonable to attribute so much importance to the condition of Europe at that time (which they seem always to have in mind). Very severe conditions are benumbing and not favourable to progress. If much progress then occurred it is likely to have been in the interglacial periods; and as well as I can judge from the scanty evidence, improvements in culture entered Europe from the south and south-east, where the influence of northern glaciation must have produced climatal changes that were stimulating without being oppressive.

§ 8. The extraordinary variability of modern Man (considered as one species) in stature, shape of skull, size and power of brain, colour, hairiness, quality of hair, and other characters, physical and mental, may be referred chiefly to his having become adapted to various local conditions upon settling here or there for long periods of time after wandering over the world in quest of game. The settling of offshoots of the original stock

[1] *Man and His Ancestors*, p. 133. [2] *Op. cit.* p. 177.

in certain regions long enough for them to undergo adaptation to local circumstances is the simplest explanation of existing races: the Negro adapted to equatorial Africa; the Asiatic stock ("Mongolian") to Central Asia; the Mediterranean race to the neighbourhood of the sea after which it is named—not that these adaptations have been made out. As to the Nordic sub-race (of the Mediterranean, we may suppose), with its fair hair and skin, it has the appearance of an Arctic beast of prey, like the Polar bear. The snow-leopard of the Himalaya is found at a midway stage of such adaptation, and the Siberian tiger is much lighter coloured than the Indian. Some geologists and zoologists now believe that, during the Glacial Period, the climate of Northern Europe was not everywhere such as necessarily to destroy the local fauna and flora, and in that case our ancestors may for ages have maintained themselves there; or, if that was impossible (as the absence of palæolithic remains in Scandinavia seems to indicate), they may have roamed for many ages along the borders of glaciation, perhaps as far as the Pacific Coast. Chinese annals refer to fair tribes in Eastern Siberia 200 years before the Christian era[1]; and it seems requisite to imagine some extensive reservoir of mankind in order to explain the origin of the vast hordes which in pre-historic as in historical times again and again invaded Europe —hordes

> "which the populous North
> Poured ever from her frozen loins, to pass
> Rhene or the Danau; when her barbarous sons
> Came like a deluge on the South, and spread
> Beneath Gibraltar to the Libyan sands."

That the race was formerly fairer than it is now may be inferred from the whiteness of its children's hair: the trait has outlived its utility. The occurrence of a fair complexion in some mountain tribes, in the Alps, *e.g.*, has occasioned the conjecture that it may be due in some way to mountainous conditions[2], of which snow might be one; but, if we suppose that the Nordic race extended during the Glacial Period into Western Europe

[1] M. A. Czaplicka, *My Siberian Year*, p. 230.
[2] Ripley, *The Races of Europe*, pp. 76–7.

(having already acquired its distinctive characters), a fair complexion in the Alps may be understood by supposing that, whilst the greater number of them followed the ice-sheet back to the north-east, some followed it southward up into the mountains—if the complexion is really ancient there. One may hazard the less serious suggestion that some of them on some occasion crossed to Africa and left their mark there in the complexion of the Berbers.

Two objections to this hypothesis will occur to every one: (i) Why are not the Esquimo fair? Because, I suppose, they are much more recent immigrants into the Arctic regions, and perhaps were fully clothed when they arrived there. (ii) Could the Nordic people have existed in such circumstances unclothed? Whether this was possible physiologists must judge. We see the Fuegians maintain themselves, practically naked, under very inclement conditions. And it is not necessary to assume that the Nordic hunters were entirely naked; since the correlation between the hair, eyes and all parts of the skin is such that, if the whitening of any part (say the hair) was sufficiently advantageous to determine natural selection, the remainder of the body would be similarly affected. And, no doubt, the Mediterranean race was always whitish.

The Amerinds seem to have been derived chiefly from the Asiatic race. Tasmanians and Australians may represent separate and still older stocks. But, as a result of migrations and conquests, most peoples are of mixed descent; and hence (i) individuals in the same locality sometimes vary greatly, because they inherit the blood of different strains in different proportions; and (ii) classification is difficult, so that whilst some observers are content to find half a dozen races, Deniker enumerates twenty-nine.

Besides general racial differences, there exist within each race and within each national group further differences between individuals in their physical, and still more in their mental, stature and ability. As it was necessary that Man should vary greatly in undergoing adaptation to the hunting life (as well as to different local environments) he was in an organic condition favourable to further variation[1]. And this has been utilized in adapting him to a certain special condition (as I believe) of his gregariousness, namely, life in the hunting-pack; for this requires some difference of personality between leaders and followers, first in the chase and later in war; and it also requires greater variability among the males, who supply the leaders, than among the females, which in fact is found still to exist. A good democrat may think it would have been a better plan to make all men and women equal from the first; and I would it had been so; for then the head of the race would not have had to drag along after it such an altogether disproportionate tail: a tail so huge and unwieldy that one may doubt whether it can ever be extricated from the morass of barbarism. But in the early days of gregariousness a pack could not have held together, or have hunted efficiently, if all had been equal and each had exercised the right of private judgment. So in successful packs one led and the rest followed; as they still do, and will continue to do, of whatever kind may be the leader. And of all the structures that make up a human being the most variable is the brain: the differences between men in stature and physique are trifling compared with those in mental power. Whatever feat of strength your Samson can perform, half-a-dozen ordinary men can also accomplish; but in every generation tasks are carried out by intellectual athletes, toward which all the ordinary men in the world combining their efforts could do nothing—absolutely nothing.

[1] Darwin, *Animals and Plants under Domestication*, Pop. ed. II, p. 308.

CHAPTER IV

CULTURAL CONSEQUENCES

§ 1. The dependence of Man upon Society both for the progress of culture and for the development of personal character is now generally recognized, and it seems to me a strong recommendation of the hypothesis I am supporting concerning his origin, that it offers a good explanation of the origin and nature of human society.

Monkeys of most species, whether in the New World or in the Old, are social, living in bands of from ten to fifty or more, and may co-operate occasionally in mutual defence or in keeping watch. Baboons, indeed, are seen in herds of several hundreds; and they are credibly reported to co-operate in raiding plantations, and in defending themselves against leopards, other baboons and even human hunters[1]. Gibbons, again, are social, going in bands to the number of fifty. But the large anthropoids live only in families—the male orang being even of a somewhat solitary habit; three or four families of chimpanzees may for a time associate together; a party of gorillas is generally limited to the family; and the condition of these apes supplies the only datum from which to judge of the condition of our own ancestor before the differentiation. Man, however, is everywhere—with a few doubtful exceptions, probably degenerate—both social and co-operative; and the purpose of his co-operation at the level of the Australian or the Semang is instructive. It is not (as we might infer from our own life) in industry, but in hunting, war, or tribal ceremonies that tribesmen work together—the last no doubt of comparatively recent origin: so that not many thousand years ago there was no co-operation except in hunting and in war (which come to the same thing).

That the large anthropoids are neither gregarious nor

[1] Numerous references might be given, from which I select Hagenbeck, *Beasts and Men*, p. 63.

co-operative follows from their having no task in which co-operation would be useful, no common purpose: the males are able alone to defend themselves and their families; and when families range apart one from another through the woods their food is in better supply. But the ancestor of Man found an object for association and co-operation in the chase. Spencer indeed, says that a large carnivore, capable of killing its own prey, profits by being solitary; and this may be true where game is scarce: in the Oligocene and Miocene periods game was not scarce. Moreover, when our ape first pursued game, especially big game (not being by ancient adaptation in structure and instinct a carnivore), he may have been, and probably was, incapable of killing enough prey single-handed; and, if so, he will have profited by becoming both social and co-operative as a hunter, like the wolves and dogs. The pack was a means of increasing the supply of food per unit; and gregariousness increased by natural selection up to the limit set by utility. Hence (as will be shown at length in a later chapter) Man is in character more like a dog or a wolf than he is like any other animal.

§ 2. The society thus formed has in its development done more than anything else to promote human life; but it could have done nothing for us if it had been merely a flocking together for mutual safety without any active purpose. It is everywhere purposeful. Still, if some remote ancestor of the anthropoids in the Primate stock had been (like most monkeys) gregarious, this trait may have persisted as a latent character which, being re-aroused in Lycopithecus by the need of co-operation, facilitated and confirmed the formation of the pack.

Some development of the rudiments of speech may be confidently traced to social co-operation. The gibbon, most social, is also the most vocal of anthropoids; but having no common task in which united action is necessary, he uses his remarkable power of voice (apparently) merely to express his feelings and to keep the troop together. The chimpanzee and the gorilla enjoy probably a close and affectionate family life, but one that makes little or no demand for concerted effort. Hence their vocalization is very rudimentary. According to R. L. Garner, it is true speech: "a chimpanzee (he says) knows the meaning of the sounds he makes, and intends to convey it to some definite individual at whom he looks. But he has at command very few sounds, and those mainly expressive of natural wants[1]."

If it be urged that anthropoids do not talk because their lower jaw and tongue have not the special adaptation to speech that is found in Man, it should be considered (*a*) that if such structure had been useful to them it would have been acquired, as at some time it must have been by Man himself; and (*b*) that even without any change they might have jabbered well enough to convey a good many discriminated, objective meanings if they had needed to do so: for Man must have begun in that way; he cannot have waited for the development of physical structure before trying to talk. Sufficient intelligence is not wanting to chimpanzees; for in captivity they learn to understand a good deal that is said to them. What they wanted was a sufficient motive for persistently trying to communicate, such that those who made any progress in the art had a living advantage over others. Man had such a motive; because co-operation was necessary to him, not (as we have seen) in industry, but in hunting. In hunting, in planning and directing the hunt, speech is plainly useful: and it is better than gesture, which probably preceded, and generally accompanied it; because, as speech became independent of gesture, it could go on whilst the hands and body were otherwise employed, or where comrades could not see one another—transferring, by a very profitable division of labour, the whole business of expression to organs not otherwise needed. It may not be much more than very simple beginnings of articulate

[1] R. L. Garner, *Gorillas and Chimpanzees*, ch. vi, where mention is made of such meanings as "food," "calling to some one," "affection," "good" (said, I suppose, of food), "warning cries," "cold or discomfort," "drink," "illness," "dead": the entire vocabulary, perhaps, not more than twenty signs. The value of Garner's work is disputed.

speech that can be traced to early co-operative hunting; but in the beginning lies the whole difficulty. And the situation was particularly favourable to the beginning of language by onomatopœia, imitating the characteristic noises of different animals and of the weapons and actions employed in pursuing and slaying them.

§ 3. The intelligence and extensive knowledge (compared with anthropoids) that distinguish Man in his lowest known condition are clearly accounted for by his adoption of the hunting life. Already (as we may assume) the most intelligent of living animals, with great knowledge of the forest, he had everything to learn about the world beyond the forest as soon as he ventured into it, and everything to learn about the art of hunting. Depending chiefly upon sight and hearing, he had to learn by observation, and to remember, and to apply all and more than all that the carnivore knows and does instinctively, or learns by following its mother. He must have learned to discriminate all sorts of animals, many of them new in a strange country; their reactions to himself, manner of flight, or of attack, or defence; the spoor of each and the noises of each; its habits and haunts, where it reposed or went to drink, where to set snares or lie in wait for it. Advancing to the use of weapons, he must have adapted them to his prey; he must have discovered the best materials—wood, or stone, or bone—for making weapons, the best materials for snares, and where to find such things. He must have fixed in his mind this series: game, weapons, the making of them, materials, where found; and must have learned to attend to the items of the series in the necessary order without impatience or confusion: a task far beyond the power of any other animal.

Further, the hunting life supplied a stimulus that had formerly been wanting to our ape. There is some difficulty in comprehending why the anthropoid in his leisurely life should be as intelligent as he is; and, similarly, it seemed to Wallace that the savage has intelligence above his needs—"in his large and well-developed brain he possesses an organ quite disproportionate to his actual requirements[1]." This illusion results from our not reflecting that the first task of increasing intelligence is to deal appropriately with details in greater and greater number and variety, and that the details of their life, with both savage and anthropoid, are just what we cannot appreciate. Still, the anthropoid seems to have rather a lazy time of it: especially, he seems to have hardly any occasion for following out a purpose needing some time for its accomplishment. This powerful stimulus the hunting life applies to carnivores, above all to dogs and wolves; and in the same way it affected our ape: compelling him to combine many activities for a considerable period of time, along with his fellows, and to direct them to one end in the actual hunting, and (later) to prosecute still other activities for a longer period in preparing weapons and snares to make the hunting more effective. Add to these considerations the development of gesture and rudiments of speech, exacting intelligence for their acquisition and increasing intelligence by their attainment, and the superiority of the lowest savage to an anthropoid is sufficiently explained. Severe must have been the selection of those that were capable of such progress, and correspondingly rapid the advance and differentiation of the species[2].

§ 4. Using stones as weapons, and finding that broken stones do most damage, and breaking them for that purpose, the progressive hunter necessarily makes some sparks fly; and if these fall amongst dry leaves or grass, he may light a fire. "In making flint implements sparks would be produced; in polishing them it would not fail to be observed that they became hot; and in this way it is easy to see how the two methods of making fire may have originated[3]." But if the

[1] *Natural Selection*, p. 193.

[2] Primates of different Families have, nevertheless, been observed when in captivity to work with great persistency at any task that excited their interest. See the account of a Cebus in Romanes' *Animal Intelligence*, and of the Orang in Hornaday's *Minds and Manners of Wild Animals*, ch. viii. Persistency of purpose is not wholly due to the hunting life but directed and confirmed by it. It is an universal character of effective animals—*e.g.*, ants and spiders.

[3] Avebury, *Prehistoric Times*, 7th ed., p. 578.

production of fire by friction had been suggested by the polishing of flints, it could hardly have been discovered before the neolithic stage; whereas hearths are known of much earlier date. And it may have happened earlier whilst some one was polishing an arrow or a spear with another piece of wood: a supposition which dispenses with the long inference from a warm flint to a flaming stick[1]. It is a curious fact that to this day in Australia fire is sometimes made by rubbing a spear-thrower upon a shield[2]; but I lay no stress upon this, as if such a practice must be traditionary from the earliest discovery of the method. Either in the chipping of flints or in the polishing of spears it is far easier, and a more probable way, to learn the art of making fire than by observing that dried boughs or bamboos driven together by the wind sometimes catch fire; because those processes include the very actions which the art employs: imitation of nature is not called for. It is true that the natives of Nukufetan in the South Seas explain the discovery of fire by their having seen smoke arise from two crossed branches of a tree shaken in the wind[3]; but this, probably, is merely the speculation of some Polynesian philosopher. Volcanoes, too, have been pointed out as a possible source of fire; and, in the myth, Demeter is said to have lit her torches at the crater of Etna—an action fit for a goddess. But were such an origin of fire conceivable with savages, it would not show how they came to make it themselves. Fire at first must have excited terror. Until uses were known for fire no one could have ventured to fetch it from a volcano, nor to make it by imitating the friction of boughs in the wind. Fires were accidentally lit by man again and again, and much damage done, before he could learn (*a*) the connection of events, (*b*) the uses of fire, (*c*) purposely to produce it, (*d*) how to control it. The second and fourth of these lessons are much more difficult than the mere making of fire; they are essential, yet generally overlooked. It seems necessary to suppose a series of accidents

[1] Mr Morris has suggested both of these ways in which the making of fire may have been discovered. *Op. cit.* p. 183.

[2] Spencer and Gillen, *Northern Tribes of Central Australia*, p. 619.

[3] Turner, *Samoa*, p. 285.

at each step, in order to show the effects of fire in hardening wood, hollowing wood, cooking game, baking and (later) glazing clay, and so forth. Perhaps a prairie-fire disclosed the advantages of cooking game, and many a prairie was afterwards burnt to that end before a more economical plan was discovered. As to the effect of fire on clay, Lord Avebury observes that clay-vessels may have been invented by (1) plastering gourds or cocoanuts with clay to protect them from the fire when boiling water in them; (2) noticing the effect of fire upon the clay; (3) leaving out the vegetable part[1]. This must have been a comparatively recent discovery; though there is some evidence of pottery having been made by palæolithic Man. It is impossible to say when the art of making fire was discovered; but it was certainly known to the Mousterian culture —say 50,000 years ago: probably very much earlier, and it was made by hunters.

§ 5. Cannibalism, where it has been found amongst extant peoples, or is known to have been formerly practised, was often justified by certain magical or animistic ideas, or by the satisfaction of revenge or of emphatic triumph over an enemy, but sometimes frankly by dietetic taste. Was it an ancient and perhaps general custom? Excavations at Krapina in Croatia disclosed along with remains of the Neanderthal species, which seems to have had a habitation there, those of rhinoceros and cave-bear and of some other kind of Man; and "some of the human bones had been apparently split open: upon that slender basis the Krapina men have been suspected of cannibalism[2]." If the suspicion is valid, the practice existed, say 50,000 years ago, in one species of Man; and perhaps much earlier, if we consider how it was merely an extension of the practice of devouring game to include the slain members of a hostile pack. For as primitive Man, or his pre-human forerunners, no doubt regarded other animals as upon the same level as himself (a bear as a bear-shaped man), so he will have regarded human enemies as upon the same footing as other animals. That true

[1] *Op. cit.* p. 579. [2] Keith, *The Antiquity of Man*, p. 134.

carnivores are generally not cannibals may be put down to their more ancient and perfect adaptation to a predatory life. For them persistent cannibalism would have been too destructive; and for us it belongs (if ever it was a general practice) to the experimental stage of our history; though of course, even in recent times, under stress of famine, reversion to the practice is not unknown to civilized men. It meets in them with no insuperable repugnance.

§ 6. Man has lost the restraint of seasonal marriage, common to the anthropoids with other animals, as determined by food-supply and other conditions of infantile welfare; though, according to Professor Westermarck, traces of it may still be found in a few tribes[1]. That our domestic carnivores have also lost this wholesome restraint on passion and population points, probably, to some condition of a steadier food-supply as determining, or permitting, the change amongst ourselves. No growth of prudence, however, or habit of laying up stores can explain the steadier supply of food, since the lower savages have no prudence and no stores. On the whole the change may be attributed (1) to an omnivorous habit being more steadily gratified than one entirely frugivorous or carnivorous; (2) to our ancestors having wandered in quest of game from country to country in which the seasons varied, so that the original correspondence of birth-time with favourable conditions of infantile welfare was thrown out. There may also have been causes that kept down the numbers of the pack so as to be equivalent in seasons of scarcity to more abundant food: the hunter's life, whilst securing a richer normal diet, involved many destructive incidents. And this (by the way) was favourable to rapid selection and adaptation; though if the destruction had been great enough to counterbalance the advantages of animal food, it must have frustrated the whole adventure. If, again, the loss of seasonal marriage tended to increase the population it would have tended to compensate the destruction due to the hunting life, and at the same time

[1] *Human Marriage*, ch. ii.

to provide more opportunities of variation. The change certainly disturbs us with an incessant urge of sexual proclivities that adds greatly to the pathology of human life; and yet it is difficult to see how without it the more romantic or the nobler characters of affection could ever have been manifested.

§ 7. *Summary of Chapters III and IV.* Of the addiction of some ancestral ape to animal food, and to the life of a hunter in order to obtain it, then, the special characteristics of Man seem to be a natural consequence; or, in other words, we may infer from the facts of human structure and condition that such an ape did adopt such a life. The hypothesis is exceptionally simple and moderate. It is generally admitted that our ancestor was a large anthropoid—possibly more gregarious than others, possibly more apt to live upon the ground; but neither of these suppositions is requisite. He was probably adapted, like the chimpanzee and gorilla, to a forest life: in which they may be supposed to have gone on for ages with little change, except some increase of intelligence which they may have shared with nearly all the Mammalia of the Cainozoic era. But into that ape's life a disturbing factor entered—the impulse to attack, hunt and eat other animals, which extensively replaced his former peaceable, frugivorous habit. The cause of this change may have been a failure in the supply of his usual diet, or an accidental variation of appetite. Not a great number need have shared in the hunting impulse: it would have been enough that a few should share in it; or, theoretically, even one may have sufficed. If advantageous and inheritable it would spread through the descendants, who would also learn from the example and instruction of their parents. It was advantageous (1) in extending their resources of nutrition, and (2) in enabling them to escape from the tropical forest. On the other hand, to those least fit for the new life it brought the disadvantages of more strenuous exertion and of competition with other carnivorous types. But with hands and superior intelligence, those that had the requisite character succeeded. There was rapid selection of those whose variations of structure, character, activity were most effective in dealing with game and with enemies; especially of those who combined and co-operated, and learned to direct co-operation by some rudimentary speech.

But, again, the hunting impulse here assumed to have possessed some anthropoid was not something entirely new; anthropoids and many other Primates are known to seize and devour birds, lizards and even small mammals when chance offers an easy opportunity. It is merely a clearer purpose and a greater persistency in such behaviour that turns it into hunting. How very improbable that such a change should *not* sometimes occur! Is it not likely to have occurred often and with many failures? Similarly of the resulting changes: the differentiation of our hands and feet is only an advance on what we see in the gorilla; and as for our ground-life, can the adult male gorilla be fairly called "arboreal"? Several Primates use unwrought weapons; most of them lead a gregarious life, to which our own is a return; they are co-operative at least in defence; like many other animals, they communicate by gestures and inarticulate vocal cries. Co-operative hunting, indeed, may seem to be new in our Order; but since wolves and dogs, or their ancestors, fell in with it at some time or other, why should it be beyond the capacity of apes? The co-operative raiding of plantations by baboons is in some ways similar to hunting in pack. Thus at each occasion of change in structure or function Lycopithecus merely carried some tendency of the other Primates a little further and a little further; until, certainly, he went a long way. Granting the beginning (which I have said is obscure) the whole movement can be distinctly pictured throughout, and it has an air of being natural and even inevitable. Few hypotheses ask us to grant less than this one.

Moreover, if the story is not true, Man is an exception to the rule of animal life, that the structure of every organism is made up of apparatus subserving its peculiar conditions of nutrition and reproduction. Indeed, conditions of nutrition are the ground of the differentiation of animals and plants.

Conditions of reproduction need not here be considered, as the apparatus[1] is the same in the anthropoids as in ourselves. With many species to avoid being eaten or to mate are the reasons for some secondary characters, such as protective armour or coloration, fleetness with its correlative structures, nuptial plumage, and so forth. But to avoid being eaten and to mate, it is first of all necessary to eat and live; and accordingly, for each sort of animal, starting from the organization of some earlier stock, its structure and activities are determined by the kind of food it gets and the conditions of getting it: in our own case, the hunting of game afoot.

[1] Its functioning, however, seems to have altered. Mr Hornaday reports that, in the New York Zoological Park, in 1919, Suzette married Boma (chimpanzees), and after seven months' gestation a baby was born. This with our own infants is the shortest period compatible with survival. Again, it is reported from the same institution that with both chimpanzees and orangs "sexual maturity is attained at about six years" (Prof. R. S. Lull, *Organic Evolution*, p. 660). Whether we can be confident that in the wild state these periods are the same as in captivity I do not know; but it seems reasonable to expect them to be shorter in the anthropoids than in Man in view of his greater individual development.

CHAPTER V

MENTAL DIFFERENTIATION OF MAN

§ 1. Heredity, Adaptation, Accommodation. Following the general belief that Man is descended from a stock nearly allied to the greater anthropoids—orang, chimpanzee, gorilla—we may assume that his mental endowments were once much the same as theirs; and that, so far as they are still the same, *heredity* sufficiently explains his having them. Thus the senses, perception, the simpler forms of comparison and inference, the appetites and many of the instincts and emotions are common to us with the apes, are seen in our children under three years old, and (in short) constitute that generic consciousness (as I have called it) from which the human mind in general and the peculiar traits of races and individuals are differentiated.

So much for heredity; but the differences of the human from the anthropoid mind, alike in intelligence and in character, are enormous, and must be accounted for in some other way. Allowing for some original specific difference which we can hardly hope to discover, the changes that have taken place may be considered as the result of *adaptation* to those habits of life under which our species (now ranking zoologically as a Family) has been developed. And this adaptation I shall assume to have been brought about under conditions of natural selection (in the sense explained in Chapter I): human races, as we now see them, being the survivors of many variations, more or less successful, and the others having been destroyed. For good judges are of opinion that, amongst the discovered remains of ancient specimens of the human Family, some that exhibit marked deviations from the modern type—Neanderthalensis, Eoanthropus, Pithecanthropus—should be regarded not as belonging to our ancestral line, but rather as representing distinct species or genera that have failed in the struggle for existence. Racial mental characters can be observed with confidence only where there has not been too great miscegenation, and nowhere have they yet been clearly determined.

But besides the innate dispositions of human nature determined by heredity and natural selection, which are found in some measure universally, because they are adaptations to conditions that, at one time and not long ago, weighed upon the ancestors of all of us, there are numerous traits (some of them quite superficial) that vary from country to country and from age to age, according to geographical circumstances, the economic or political type of the society in which a man lives, his place therein, religious institutions and the countless causes that govern manners and customs. In the lives of most men these traits are not necessary; they may be adopted and cast aside more than once in an individual's career: they are temporary *accommodations* due to education, imitation, tradition; and, in fact, are often the disguises of human nature. Still, as society grows more and more complex, orderly and stable, there is, no doubt, again some natural selection of those individuals who are capable of undergoing the requisite accommodations. Those that cannot endure the restraints of civilization, wander away; the extremely lazy, improvident, dishonest, or aggressive, in considerable numbers, perish—but not nearly enough of them.

§ 2. Probable Mentality of the Original Stock. To the original mentality of man we can only seek a clue in the higher Primates, and especially in the extant anthropoids. No doubt, during the long millennia that have elapsed since the separation of our own stock from those of other genera and species, they also have undergone some evolution, but much less change than we have. Probably they, like most other mammals, have gained in intelligence. Unfortunately, our knowledge of their habits and abilities is still deplorably limited. It seems certain, however, that their intelligence is much greater than that of any other kind of animal. Professor R. M. Yerkes, as the result of experiments on a young orang (4–5 years old) concludes that, "as compared with monkeys and other mammals, the orang-utang is capable of expressing free ideas in considerable numbers and of using them in ways highly indicative of thought-processes, possibly even of the rational order. But contrasted with that of Man, the ideational life of the young orang-utang seems poverty-stricken, certainly in this respect Julius (the young orang) was not above the level of the normal three year old child[1]." Dr Köhler in his *Intelligenzprüfungen an Menschenaffen*, after describing his experiments upon chimpanzees in Madeira, expresses general agreement with this opinion. He finds that the behaviour of chimpanzees in obtaining fruit by means of sticks, boxes piled one on another and in other ways, shows insight; which I take to mean a practical understanding of the situation and of the way to gain their end. Their intelligence ranks between that of Man and that of the lower mammalia but is much nearer the latter. The great apes (he says) show many individual differences in character and intelligence. In their native state they must have extensive knowledge of their habitat, of all the forest can yield for food or shelter, and of its other denizens dangerous or otherwise. They construct for themselves some sort of sleeping-place, not much inferior to the Australians' "lean-to," by piling branches together in the trees. Toward men, anthropoids seem to be naturally unaggressive, and usually retreat from them, but when attacked, defend themselves with fury, and become suspicious and aggressive when they are much persecuted. From other animals the male gorilla has nothing to fear, and he defends his family against leopards; the chimpanzee is said to fight leopards with varying success; and, as for the orang, Dyak chiefs told Wallace that no animals dare attack him, except crocodiles and pythons, and that he kills both of them[2]. Their food (as I have said) is chiefly fruit and the tender shoots of trees and bamboos; they sometimes eat eggs, young birds, small mammals that happen to fall in their way, but do not hunt. Socially, they hardly get beyond family life. Orangs, male

[1] *The Mental Life of Monkeys and Apes*, p. 132.
[2] *Malay Archipelago*, pp. 46–7.

and female, are even seen alone, and young ones together without parents; gorillas are seen in family parties; chimpanzees in families, and occasionally three or four families in company. It is said that gorillas and chimpanzees have been seen together in a large band. I have met with no report of these animals fighting amongst themselves, except that male gorillas sometimes fight for a wife, and that may be true of males of the other species. Gorillas have also been said, upon very slight evidence, to be polygamous; chimpanzees and orangs seem to be monogamous[1]. Their family life is probably, as amongst all the other Primates, affectionate: the long youth of their children implies much parental care. Whilst the smaller anthropoids—siamang and gibbon—go in troops, as also do the baboons and most monkeys of both hemispheres, the less sociability of the great anthropoids may be understood to result (as I have said) from the limited supply of the right sort of food for them, even in the tropical forest to which they are confined—since animals of their bulk must consume a great deal; and from their having no need of combining for the purpose of defence.

From the type thus outlined the mentality of the human race has departed so widely that some even of those who believe that our bodies have been derived from some simian stock hesitate to admit that our minds can have had a similar history. But as everywhere else in the animal kingdom mind and body constitute one organism, it is reasonable to consider whether the differentiation of the mind of man may not be understood to have taken place under the same conditions as those which determined the transformation of his body. What were those conditions?

§ 3. In the foregoing chapters I have collected a number of facts and arguments pointing to the probability that (*a*) the chief cause of the evolution of the human Family was the adoption by some anthropoid (or allied form) of the life of the hunter

[1] According to R. L. Garner, however, both gorillas and chimpanzees are polygamous. See *Gorillas and Chimpanzees*, pp. 54, 214.

in order to obtain animal food. That the change from a frugivorous to a carnivorous diet may itself have had some effect upon our temperament and activity is probable; but I lay no stress upon that. Most monkeys are almost exclusively frugivorous; langurs even prefer leaves to fruit, and their stomachs are adapted to this coarser diet; yet monkeys are the most alert and active of animals; some of them are amongst the most courageous; anthropoids are amongst the most powerful. A carnivorous diet alone would not explain any changes in the shape and proportions of our trunk and limbs, nor the upright gait, nor the gregarious habit, nor the development of the brain, nor the invention of weapons, nor any of the mental and emotional characteristics that distinguish man from the other Primates; but all these things readily follow from our remote ancestor's adoption of the life of the hunter.

Sociologists, surveying extant peoples, have usually distinguished four stages of culture, the hunting, pastoral, agricultural and manufacturing; and some have indicated what they suppose to have been a still earlier stage, the "collecting," such as may be seen amongst the Fuegians and has been practised upon kitchen-middens all the world over. But 'collecting,' as the chief livelihood, is the degenerate resource of tribes fallen into distress; it cannot have been the first stage, because it implies no conditions that tend in any way to develop body or mind or society. That hunting came first is a true intuition, and to understand the development of human nature we have only to refer the hunting-life back to the very origin of the human stock[1]. Mr A. F. Shand has pointed out[2] that this is in agreement with Leplay's doctrine that the work we do and the life we lead are the chief causes of a change of character.

(*b*) With the dispersal of our Family throughout the world,

[1] This view is not opposed to the suggestion I have somewhere seen that the collecting activities of women, whilst men hunted, may at some stage have led to private property and to the domestication of plants and animals. Again, the pastoral and agricultural states are not necessarily successive. It depends on local conditions. For an excellent survey of the gradual rise of primitive culture and the difficulties it encountered, see H. Spencer's *Industrial Institutions* (*Princ. of Soc.*, III).

[2] *Sociological Review*, July, 1921.

geographical conditions operated both stimulatively and selectively. The great anthropoids are all confined to the equatorial forest, and it is obvious that, with their diet, it is impossible to pass out of tropical or (at furthest) sub-tropical regions. But the adoption of a flesh diet, having enabled the human stock to extend the range of its hunting (allowing for gradual adaptation to climate or accommodation by clothing) to any country that supplied the requisite prey, the settling of various offshoots of the original stock in certain regions long enough for them to undergo adaptation to local conditions, if it offers the simplest explanation of existing races physically, may also be considered as an important determinant of whatever mental and moral differences may be assigned to them.

(*c*) Whilst none of the great anthropoids has advanced socially beyond family life, man is everywhere (with few and doubtful exceptions) gregarious—living at the lowest grade in tribes or bands of about fifty; and the gregarious life is the most influential of all the conditions of his cultural development. Possibly, he may have been originally more gregarious than any of the great extant anthropoids, in spite of his not having needed society for defence and of its seeming to be (for so large a frugivorous animal) inconvenient in relation to nutrition. Moreover (as I have suggested), if the greater anthropoids and our own ancestors were descended from some stock of the lower monkeys, such as always go in troops, the gregarious instinct, even if it had ceased to be active, may have remained with them as a latent character. Still, it is my conjecture that Man became gregarious, or recovered the social habit, because of the utility of co-operative hunting, so that he became at first a sort of wolf-ape.

This hypothesis helps us to understand (1) why every human society is still formed upon the type of the hunting-pack, and (2) why Man is still so imperfectly sociable: the purpose of the hunting-pack, each wolf-ape seeking prey, was unfavourable to social life in other relations. That in human life group-consciousness preceded self-consciousness is a groundless and fantastic notion: all known savages are fully self-conscious, as their sentiments and behaviour imply, and even the higher brutes are (in my judgment) self-conscious in their relations with others. Current speculations about fashion, imitation, tradition, crowd-psychology, are in danger of exaggeration, and overlook the patent facts of individualism as shown both by the general insistence upon private property and personal liberty, and also by the multitude of eccentrics—the hypocrite, the criminal, the miser, the vagrant, the contra-suggestible, the hermit, the sceptic, the saint. Some people, without being in any way morbid, find that a good deal of solitude is necessary to the complete life; by nature the student and pioneer escape from the crowd. One way in which the imperfect socialization of Man shows itself—the need of external opposition in order to give internal unity—is the tendency of every large aggregate to split up into mutually hostile sections, "nationalities," classes, parties, trades, each with the spirit of the ancient hunting-pack. "To cantonize is natural," says Shaftsbury (*Sensus Communis*), "when the society grows vast and bulky ...vast empires are in many respects 'unnatural'." The cantons proceed to make war one on another. Great states may, indeed, long be preserved by their military power and by economic advantages; but they are compounds and rarely well-co-ordinated except as a pack in war.

(*d*) The later stages of human development have been essentially modified by certain imaginary conditions peculiar to Man; for he—we know not at what date—invented them. These may be summed up under the names of Magic and Animism; and in *Man and His Superstitions* I have discussed them with their astonishing vagaries and still more astonishing reactions upon human life. Man, wherever we find him, is always subject to social regulation, and this regulation is always maintained in large measure by supernatural beliefs to which his mentality is accommodated, or perhaps adapted, so that it may not be easy to live without them.

The chief conditions, then, to which mankind has been adapted, and thereby differentiated in body and mind from the anthropoid stock, I take to be four: the hunting life; geographical circumstances; social life; and his own imaginations.

§ 4. Primal Society. In looking for the probable form of the earliest human or (rather) pre-human society, one naturally makes a survey of other mammalian societies; and the task is soon accomplished. It is surprising how few and simple the types of them are, in contrast with the elaborate polities of some hymenoptera and of the termites: these have much greater superficial resemblance to modern human societies; but in fact, they are families rather than societies; their interesting activities will one day probably be traced to relatively simple mechanisms; and in every way they are too remote from us for any useful comparison. As for mammalian societies, even using the term to include families, they may be classified under four or five types:

(1) Families: (*a*) Monogamous: of which the best examples seem to be found in some monkeys. Many of the cats are believed to pair monogamously; but it is doubtful whether, or in what measure, the male takes part in the rearing of the whelps.

(*b*) Polygamous: characteristic of many species of deer; after the breeding-season, the stags often wander away by themselves.

(2) Associations of families without apparent structure or organization, such as those of the vizcacha and the beaver. They have no leaders, and make no attempt at mutual defence; but their inco-ordinated activities, in making their burrows, dams, etc., have results which, especially in the case of the beavers, look as if the animals had worked upon a common, premeditated plan. Gregariousness exists widely in the animal kingdom without any utility in attack or defence, but merely for convenience of breeding, or for the advantage of signalling the approach of danger, from any direction, to the whole flock.

(3) Troops or herds, comprising several families. This type is common amongst monkeys: generally the families are

monogamous, and both parents care for the offspring; there are leaders, and the members sometimes combine in mutual defence. This organization is especially effective with the baboons—who, however, are polygamous. A somewhat similar type is characteristic of cattle who also have leaders as the result of battle between the bulls, each trying to control and keep together as many cows as he can; and they often combine their forces against beasts of prey. A herd of elephants, however, is often led by an old cow, who sets a pace practicable for the calves; and red deer, when they travel in a herd, are led by an old hind—apparently upon general consent as to her superior sagacity.

(4) Hunting-packs—most noticeable with wolves and wild dogs; they have leaders, and probably an order of precedence determined by battle. In the breeding-season (February to August) a pack of wolves breaks up into pairs; but whether their pairing is for life or merely seasonal is disputed; and it is also doubtful whether the male takes any share in caring for the puppies; such habits may vary in different localities[1]. The numbers of the pack depend on circumstances, are now much smaller than formerly in Canada and probably greatest in Russia.

Was our own primitive society, then, like any of these? Since direct evidence cannot be obtained, we must be guided in forming our judgment by two considerations: (*a*) what type of society gives the best explanation of human nature as we now find it? and (*b*) for which type can we give the best reason why it should have been adopted? So I point out (*a*) that Man, in character, is more like a wolf or dog than he is like any other animal; and (*b*) that for the forming of a pack there was a clear

[1] It is certainly believed by fox-hunters that a fox feeds his vixen when she is occupied with their family, and that "if the vixen is killed he will bring up the family by himself"—(Thomas F. Dale, *The Fox*, pp. 12, 13). Nothing incredible in this—nor of wolves. Can the vixen provide for herself and litter alone? If not, the dog must do it: else there could be no foxes or wolves. However, de Canteleu denies that the he-wolf takes any part in rearing the young (*La Chasse du Loup*, p. 30); whilst W. T. Hornaday (*Minds and Manners of Wild Animals*, p. 223), who calls the grey wolf "the most unmoral animal on earth," says that his one redeeming trait is that he helps to rear the pups—after they have been successfully defended against him by their mother.

ground in the advantage to be obtained by co-operative hunting[1].

It must be admitted that Darwin, discussing sexual selection in man, seems to suggest a different hypothesis. He says: "Looking far enough back in the stream of time, and judging from the social habits of man as he now exists, the most probable view is that he aboriginally lived in small communities, each with a single wife, or if powerful with several, whom he jealously guarded against all other men. Or he may not have been a social animal, and yet have lived with several wives, like the gorilla; for all the natives 'agree that but one adult male is seen in a band; when the young male grows up, a contest takes place for mastery, and the strongest, by killing and driving out the others, establishes himself as the head of the community.' The younger males, being thus expelled and wandering about, would, when at last successful in finding a partner, prevent too close interbreeding within the limits of the same family[2]." The information concerning the polygamy of the gorilla, quoted here from Dr Savage, who wrote in 1845, has not since (I believe) been confirmed, except by Professor Garner[3].

Naturally, the above passage has attracted the attention of anthropologists; but we must observe that it is concerned not with the pre-human but with a primitive human state of society, and that there is no suggestion that the "small communities" were not hunters. Atkinson in his essay on *Primal Law*, edited with qualified approval by Andrew Lang, starts from Darwin's hypothesis, and modifies it by not noticing the "small communities" and by urging that the young males, when driven off by their father, did not wander away, but kept near the family, always on the watch to murder their father. This amendment he makes, because he had observed the same habits in cattle and horses. Then through a row of hypotheses

[1] W. P. Pycraft, in his entertaining *Courtship of Animals*, after assuming that Man became a hunter for the sake of the excitement such a life afforded, goes on (p. 23): "A little later the advantages of neighbourliness were borne in on him, largely for the sake of the greater ease wherewith the animals of the chase could be captured by their combined efforts; but this begat comradeship and some of the graces that follow therefrom."

[2] *Descent of Man*, ch. xx.

[3] See above, note on p. 45.

with little evidence or rational connection, he arrives at an explanation of certain savage laws of avoidance, exogamy, etc. More recently, Professor Freud has produced a most ingenious and entertaining essay on *Totem und Tabu*, in which he builds upon the same foundations. You easily see how the "Œdipus complex" emerges from such a primitive state of things, but will hardly, without reading the work, imagine the wealth of speculation it contains or its literary attractiveness. It is, however, based upon inadequate studies. Atkinson probably relied upon the supposed parallel case of wild cattle and horses, because those animals resemble the apes and also (as he seems to assume) primitive man, in being vegetarian: though the diets are, in fact, very different. But even if such a comparison indicated a possible social state of our original ape-like stock, what is there in such a state that can be supposed to have introduced the changes that made our forebears no longer ape-like? Supposing those changes to have already taken place, what evidence is there that the same social state endured? None: for it was assumed to have been the social state of our forebears on the ground of their resemblance in diet and family economy to the gorilla.

§ 5. Returning, then, to our hypothesis as to the chief cause of human differentiation, namely, that a certain Primate, more nearly allied to the anthropoids than to any other, became carnivorous and adopted the life of a hunter, there are (as I have said) two ways in which this may have happened: either by such a variation on the part of our ancestor that he felt a stronger appetite for animal food than the gorilla does (strong enough to make him hunt for prey), or by such a change of climate in the region he inhabited—say from sub-tropical to temperate—as to make his former diet scarce, especially in winter, so that he became a hunter to avoid starvation. Every one admits that he became a hunter at some time: why not at the earliest? Nothing less than some great change of life, concentrating all his powers and straining every faculty, can possibly account for the enormous differentiation of Man. The adoption of the hunting life is such a change, and the further back we put it, the better it explains the other changes that have occurred in our physical and mental nature.

From an early date, again, our ancestor may have attacked big game, probably Ungulates—to whom he owed much; for not only did they provide prey, but by clearing the forest over wide areas compelled him to run in pursuit remote from his native trees, thus giving great selective advantage to every variation of legs and feet adapted to running: though at the very first there may have been little need to run, as he was not yet an object of terror; "we must remember that if man was unskilful, animals were unsuspicious[1]." I suppose him, at first, to have fallen to with hands and teeth: combining with others in a hungry, savage onslaught. By attacking big game, advantage was given to those individuals and families who co-operated in hunting; thus forming the primal society of the pre-human stock; a society entirely different from that of any of the Primates, or of cattle, and most like that of the dogs and wolves—a hunting-pack.

As in the course of generations the hunting-pack developed, no doubt it had recognized leaders, the most powerful males, one perhaps pre-eminent. But it was not subject to one old male who claimed all the females (as Atkinson supposed), for the more adult males it comprised, the stronger it was; and for the same reason, pairing, as among wolves, was the most efficient form of sexual relationship. But, in my judgment, it is altogether vain to try to deduce from this form of society, which may have existed three or four million years ago, any of the known customs of savages concerning marriage, such as avoidance, totemism, exogamy; which would be of comparatively recent date if we put back their origin 50,000 years. Many such rules can only have arisen when there was already a tradition and a language capable of expressing relationships.

The general structure of any human society, however, whether of the tribe or nation or of some subordinate group —a trades-union or a music-society—retains the original character of the pack—a common object of pursuit and an organization of leaders and followers.

[1] Avebury, *Prehistoric Times*, 7th ed., p. 580.

CHAPTER VI

PSYCHOLOGY OF THE HUNTING-PACK

§ 1. Possibly our ape-like ancestors were more sociable than any of the extant anthropoids; but sociability in ape-life would in no way account for our present character as men, nor for the structure of our societies; nothing accounts for these things except the early formation of the hunting-pack. Since, however, we can know nothing of that institution directly, we must try to learn something about it from the parallel case of dogs and wolves. Galton remarks how readily the proceedings of man and dog "are intelligible to one another. Every whine or bark of the dog, each of its fawning, savage, or timorous movements is the exact counterpart of what would have been the man's behaviour, had he felt similar emotions. As the man understands the thoughts of the dog, so the dog understands the thoughts of the man, by attending to his natural voice, his countenance, his actions[1]." No more, if as much, could be said of the terms upon which we stand with a tame chimpanzee, in spite of greater physical and facial resemblance and nearer kinship. What can connect us so closely in mind with an animal so remote from us in lineage and anatomy as the dog is? Adaptation to the same social conditions, the life of the hunting-pack.

§ 2. (1) The master-interest of every member of the pack lies in the chase, because success in the chase is necessary to life. To show how this passion actuates ourselves, I quote Mr F. C. Selous; who, during an expedition in Canada, roused a caribou stag within twenty yards, saw "the dreadful terror" in his eyes, and shot him. "Did I feel sorry for what I had done? it may be asked. Well! no, I did not. Ten thousand years of superficial and unsatisfying civilization have not

[1] *Inquiry into Human Faculty*, p. 262.

altered the fundamental nature of man, and the successful hunter of to-day becomes a primeval savage, remorseless, triumphant, full of a wild, exultant joy, which none but those who have lived in the wilderness, and depended on their success as hunters for their daily food, can ever know or comprehend[1]." To the hunter my paradox must seem a truism. And that the hunter temporarily released from civilized restraints, who suffers such intoxication, merely renews old savage raptures is shown by the following curious parallel: a Boschman, returning from a successful hunt to the wagons of the traveller Baines—"Behold me!" he shouted, "the hunter! Yea, look on me, the killer of elephants and mighty bulls! Behold me, the big elephant, the lion! Look on me, ye Damaras and Makalaka; admire, and confess that I am a great Bull-calf[2]."

Again, since the interest of the chase culminates in the kill —for this is the condition of making a meal—to kill becomes, in some predatory animals, a passion that is often gratified without regard to their needs. Wolves often slay many more sheep than they devour: a sheep-dog that undergoes reversion kills by night the sheep on neighbouring farms without any call of hunger; and, says Mr Thompson Seton (writing of the natives of North Canada), "the mania for killing that is seen in so many white men, is evidently a relic of savagery; for all these Indians and half-breeds are full of it[3]." They fired at everything they saw. The manners of my own pack—now long dispersed—were very similar to the Indians'; and the sport of pigeon- or of pheasant-shooting has been reduced to its last element—skilful slaying.

The disposition to slay is reinforced, when prey makes serious resistance, by anger; and generally by a distinct tendency, sometimes called "destructiveness," perhaps a latent character derived from the monkeys, and which I take to be in them partly a play-impulse and partly an expression

[1] *Hunting Trips in North America*, p. 349.
[2] G. W. Stone, *Native Races of South Africa*, p. 91.
[3] *The Arctic Prairies*, p. 20.

of curiosity. Some carnivorous animals (*e.g.* tigers), as they advance in life, are said not to kill more than they need.

(2) The gregariousness of the pack is variable; probably, amongst wolves, it was much greater anciently than it is to-day. There are conflicting statements about the gregariousness of wolves that have been studied in different countries. Couteulx de Canteleu (France) says: "The wolf is an enemy of all society; when they assemble it is not a pacific society, but a band of brigands[1]." Thompson Seton (Canada) says: "Wolves are the most sociable of beasts of prey; they arrange to render one another assistance. A pack seems to be an association of personal acquaintances, and would resent the presence of a total stranger[2]." Gregariousness of wolves must be reduced by failure of game (as by the destruction of bison in North America), and still more by the encroachments of civilization (as in France). The primitive pre-human pack, probably, was more constantly gregarious than wolves are: (*a*) because its individuals, having at first no instinctive or traditionary knowledge of hunting, were more dependent on co-operation; and (*b*) because the long youth of children made it necessary for parents to associate with the pack during their nurture—else no pack could have existed; for whilst wolves are nearly full-grown at eighteen months, apes are not mature until the seventh or eighth year. At a later period, after the invention of effective weapons, an individual became, for many kinds of game, less dependent on co-operation; but by that time, the hunting-grounds of a pack were circumscribed by those of other hostile packs; so that no one dared go far alone.

(3) With gregariousness went, of course, (*a*) perceptive sympathy—every animal read instantly in the behaviour of others their feelings and impulses; (*b*) contagious sympathy—the impulses of any animal, expressed in its behaviour, spread rapidly to all the rest; and (*c*) effective sympathy, so far (at least) as that all united to defend any associate against

[1] *La Chasse du Loup*, p. 21.
[2] *Life Histories of Northern Animals*, p. 755.

aggression from outside the pack. Perceptive and contagious sympathy, however, extend beyond the limits of the pack or the species. Most of the higher mammalia can read the state of mind of others, though of widely different kinds, in their expression and behaviour; and many are liable to have their actions immediately affected by signs of the emotional impulses of others, especially of fear—as may be seen in the fortuitous asemblages of different species often met with in Africa. These modes of sympathy, therefore, though liveliest amongst gregarious animals, are not dependent on specific gregariousness.

(4) The pack has a disposition to aggression upon every sort of animal outside the pack, either as prey or as a competitor for prey: limited no doubt by what we should call considerations of prudence or utility; which must vary with the size of the pack, the prowess of its individuals, the possession of weapons, and so forth. After the invention of weapons and snares, many savage tribes can kill every sort of animal in their habitat, as the palæolithic Europeans did some thousands of years ago. From the outset the pre-human pack must have come into competition with the true carnivores, must have defended itself against them, may have discovered that attack was the safest defence, and may have been victorious even without weapons. Mr G. P. Sanderson writes: "It is universally believed by the natives (of South India) that the tiger is occasionally killed by packs of wild dogs.... From what I have seen of their style of hunting, and of their power of tearing and lacerating, I think there can be no doubt of their ability to kill a tiger....Causes of hostility may occasionally arise between the tiger and wild dogs through attempted interference with each other's prey[1]."

(5) A hunting-pack, probably, always claims a certain territory. This is one of the first grounds of the sense of property, so strongly shown by domestic dogs: the territorial

[1] *Wild Beasts of India*, pp. 275–6. Cf. Casserly, *Life on an Indian Outpost*, pp. 94–5. Brehm says, in *Thierleben*, that in Russia wolves attack and kill the bear.

claims of the half-wild dogs of Constantinople are well known. To nourish a pack the hunting-grounds must be extensive. Mr Thompson Seton says that in Canada the wolf has a permanent home-district and a range of about fifty miles[1]. Very many generations must have elapsed before the deviation of our forebears from anthropoid habits resulted in the formation of so many packs as to necessitate the practical delimitation of hunting-grounds. Then the aggressiveness of the pack turned upon strangers of its own species; the first wars arose, and perhaps cannibalism on the part of the victors. It is certain that, in North America, wolves kill and eat foxes, dogs, coyotes; and it is generally reported both there and in Russia that wolves will eat a disabled companion. Mr W. T. Hornaday confirms this report[2].

(6) A pack must have a leader, and must devotedly follow him as long as he is manifestly the best of the pack; and here we have a rudimentary loyalty. The following of the leader of a pack in pursuit of prey is not mere imitation: it involves the belief that he knows where the prey is, or can at least find it. In a pack of hounds it is not every dog that by giving tongue can obtain a following: some are trusted, and others disregarded.

(7) Every individual must be subservient to the pack, as long as it works together; and this seems to be the ground of the "instinct of self-abasement" (McDougall), so far as the attitudes involved in such subserviency are due to a distinct emotional impulse, and are not rather expressive of fear or of devotion.

(8) The members of the pack must be full of emulation; in order that, when the present leader fails, others may be ready to take his place.

(9) For the internal cohesion of the pack, there must be the equivalent of a recognized table of precedence amongst its members; and this is reconciled with the spirit of emulation, by fighting until each knows his place, followed by complete submission on the part of the inferior. Mr Th. Roosevelt says

[1] *Life Histories of Northern Animals*, p. 754. [2] *Op. cit.* p. 223.

of a pack of dogs employed in bear-hunting, "at feeding-time each took whatever his strength permitted, and each paid abject deference to whichever animal was his known superior in prowess[1]." Mr W. H. Hudson writes of dogs on cattle-breeding establishments on the pampas, that he presumes "they are very much like feral dogs and wolves in their habits. Their quarrels are incessant; but when a fight begins the head of the pack, as a rule, rushes to the spot," and tries to part the combatants—not always successfully. "But from the foremost in strength and power down to the weakest there is a gradation of authority; each one knows just how far he can go, which companion he can bully when in a bad temper or wishing to assert himself, and to which he must humbly yield in his turn[2]." The situation reminds one of a houseful of schoolboys, and of how ontogeny repeats phylogeny. Where political control is very feeble, as in mining camps or backwoods settlements, civilized men revert to the same conditions. Fifty years ago, "all along the frontier between Canada and the United States, every one knew whom he could lick, and who could lick him[3]." Amongst Australian aborigines, we are told that "precedence counts for very much[4]."

(10) A pack of wolves relies not merely upon running down its prey, but resorts to various stratagems to secure it: as by surrounding it; heading it off from cover; driving it over a precipice; arranging relays of pursuers, who take up the chase when the first begin to flag; setting some to lie in ambush while the rest drive the prey in their direction. Such devices imply intelligent co-operation, some means of communicating ideas, patience and self-control in the interests of the pack and perseverance in carrying out a plan. Failure to co-operate effectually is said to be punished with death. The progenitors of man, beginning with more brains than a wolf, may be supposed soon to have discovered such arts and to have improved upon them.

[1] *Outdoor Pastimes of an American Hunter*, p. 70.
[2] *The Naturalist in La Plata*, pp. 336–7.
[3] Hiram S. Maxim, *My Life*, p. 57.
[4] Spencer and Gillen, *Across Australia*, p. 388.

(11) When prey has been killed by a pack of wolves, there follows a greedy struggle over the carcass, each trying to get as big a meal as possible. Mr Th. Roosevelt writes of dogs used in hunting the cougar (puma): "The relations of the pack amongst themselves (when feeding) were those of wild beast selfishness....They would all unite in the chase and the fierce struggle which usually closed it. But the instant the quarry was killed, each dog resumed his normal attitude of greedy anger or greedy fear toward the others[1]." As this was a scratch pack of hounds, however, we cannot perhaps infer that a naturally formed pack of wolves is equally discordant, or that the human or pre-human pack was ever normally like that. Galton, indeed, says: "Many savages are so unamiable and morose as to have hardly any object in associating together, besides that of mutual support[2]"; but this is by no means true of all savages. I agree with Mr F. C. Bartlett that "primitive comradeship is socially a very important attitude among savage peoples." At any rate, the steadier supply of food obtained by our race since the adoption of pastoral or agricultural economy, with other circumstances, has greatly modified the greedy and morose attitude in many men and disguised it in others; though it reappears under conditions of extreme social dislocation, and it is a proverb that "thieves quarrel over their plunder."

In the original pack a struggle over the prey may have subserved the important utility of eliminating the weak, and of raising the average strength and ferocity. But some custom must have been established for feeding the women and children. No doubt when fruits were obtainable, the women and children largely subsisted upon them. But the strong instinct of parental care in the Primates, the long youth of children, and the greater relative inferiority of females to males (common to anthropoids and savages) than is found amongst dogs and wolves, must have made the human pack from the first differ in many ways from a pack of wolves.

So much, then, as to the traits of character established in

[1] *Op. cit.* pp. 6–7. [2] *Op. cit.* p. 78.

primitive man by his having resorted to co-operative hunting: they all plainly persist in ourselves, and are entirely different from those of the anthropoids, who are neither aggressive hunters nor gregarious, and are therefore wanting in the traits that depend on social life.

§ 3. On our intelligence life in the hunting-pack had an influence no less revolutionary than on our character, as already explained in the fourth chapter. The whole art of hunting had to be learned from its rudiments by this enterprising family. With them there was no inherited instinct or disposition to hunt, and no tradition or instruction, as there is with the true carnivores whose cubs or whelps learn the craft by following their mothers: they depended solely on observation, memory, inference. With poor olfactory sense (as usual with apes) prey must be followed and inconvenient enemies outwitted, by acquiring a knowledge of their footprints and other visible signs of neighbourhood, and by discrimination of all the noises they make. The habits and manners of prey and enemies, their hours of activity and rest, their favourite lairs and feeding-grounds and watering-places, their paths through the forest, marsh, thicket and high grass, must all be learnt: so must their speed, endurance, means and methods of attack and defence. The whole country within the range of the pack must be known, its resources and its difficulties; and whenever new territory was entered, new lessons in all these matters had to be learned. This must have entailed a rapid natural selection of brains. Only a rapidly developing, plastic brain could have been capable of the requisite accommodation of behaviour in such conditions: a mechanism was required by which more and more new lines of specialized reaction were related to numerous newly observed and discriminated facts.

Besides the necessity of intelligent adjustment to external conditions physical and zoological, life in the pack was itself a demand for new modifications of behaviour. We have just seen what effect this must have had in altering the character of an anthropoid, and the change must have come about by new responses to facts newly observed or inferred and remembered. A great part of the mental furniture of every social animal consists in its knowledge of the dispositions and reactions of its fellows. Growing up, as we do, in settled society, this knowledge is assimilated gradually, becomes almost subconscious and a matter of course; but at first it must have been learned slowly and with difficulty.

The very crudest weapons, again, may be handled with variable dexterity; the best handling must be discovered and practised; and this had a high selective value for the hands as well as for the brain. Probably crude weapons were very early used; for some monkeys (and baboons generally) throw sticks or stones, or roll stones down upon an enemy. In Borneo, Wallace came upon a female orang who, "as soon as she saw us, began breaking off branches and the great spiny fruits (of the durian) with every appearance of rage, causing such a shower of missiles as effectually kept us from approaching too near the tree. The habit of throwing down branches has been doubted; but I have, as here narrated, observed it myself on three separate occasions[1]." The importance of the observation consists in its proving the existence in an anthropoid of the impulse to use missiles under the occasional stress of anger; so that it might be expected rapidly to develop under the constant pressure of hunger. The use of clubs and stones induced the discrimination of the best materials for such weapons, and where they could be found; and, in process of time, brought in a rough shaping of them, the better to serve their purposes. Then came the invention of snares and pit-falls and the discovery of poisons.

Thus the primitive human, or pre-human mind, was active in many new directions; and depending for its skill, not upon instinct or imitation, but upon observation and memory and inference, it was necessary for it to arrange ideas in a definite order before acting upon them, as in making weapons or planning a hunt, and to keep steadily before the mind a remote

[1] *Malay Archipelago*, p. 43.

purpose which was to be realized step by step; indefiniteness or confusion in such matters was fatal.

The contrast between growing memory of the past and present experience, between practical ideas and the actions realizing them that had been suspended until the right moment came, furthered the differentiation of self-consciousness amidst the world; the contrasts of co-operation and greed, of emulation and loyalty and submission, of honour and shame, furthered the differentiation of self-consciousness amidst the tribe.

§ 4. If it be asked—how much of all this development attributed to the hunting-pack might have been brought about just as well by the formation of a defensive herd, such as we see in cattle and horses?—a definite answer can be given. The herd is, of course, marked by (2) gregariousness; but the disposition that brings its members together is entirely different from that which collects the pack. It is not to seek food; for that they might do as well (or better) alone; still less is it to attack other animals. And the disposition that keeps the herd together is different from that which unites the pack: it is, in great measure, discomfort in being alone, because that is a condition of danger. To be with the herd, to smell it, to hear it feeding and moving, to rub shoulders with the rest is a profound satisfaction. Whereas amongst wolves the pack seasonally breaks up, and its individuals stray away, alone or in pairs, and are not afraid to get their own living. Lycopithecus, as I have said, was probably less independent of the pack, but not to the extent of fearing to be alone. At no period of our evolution need the solitary hunter have been helpless, especially after the invention of even crude weapons. It is true, however, that in recent times sociality has so greatly increased amongst us (or some of us) that some people seem to be uncomfortable when out of the crowd: their gregariousness has acquired something of the herd-like quality.

(3) Perceptive and contagious sympathy are of course developed in the herd; and sometimes effective sympathy in

common defence, as amongst the musk-oxen of the Arctic, and in several species of buffalo and other cattle, who at the approach of beasts of prey form a phalanx with the adult bulls and cows outside and the calves within, which no carnivore dare attack. On the other hand, they are apt to worry and destroy sick or wounded animals of their own herd, without the wolf's excuse in eating them. This has been attributed to their having mistaken the injured animal for the enemy that did the injury (a singular want of intelligence), or to an instinct to rid the herd of an encumbrance whose presence might endanger the rest.

(7) A herd that travels has leaders; but in some cases at least the recognition of them is determined by a principle the opposite of that which influences a hunting-pack, namely, safety or self-conservation. They travel to find pasture, or to reach water, or to escape from enemies; and it is sagacity and prudence, not ferocity, that makes one leader more eligible than another.

Again, there is in a herd (8) emulation amongst the males for possession of the females, and apparently some regard for (9) precedence, the ground of which is obscure: it is well-known that cows going to, or returning from, pasture observe a certain order of march.

On the other hand, of course, the herd has no (1) interest in the chase or in killing, no (4) aggressiveness when unmolested, nor (10) strategy and perseverance in attack, nor (11) greedy quarrelsomeness in feeding; for its food lies widespread, unlike the dead game from which each member of a pack must rend off as much as he can. Moreover, reliance on the herd for safety is unfavourable to individual intelligence. The conditions of life in the herd and in the pack are entirely different. Herd-life does not involve the great and decisive change which is implied in the evolution of human nature. We may conceive, therefore, of the primitive pre-human mind as a sort of chimpanzee mind adapted to the wolfish conditions of the hunting-pack. Wolves themselves have undergone no great development compared (say) with cats, for want of hands and other physical advantages which we had to begin with. If some species of baboon had taken to the hunting-life, there might have been interesting results.

I cannot help regretting that Mr Trotter should have called his excellent book *The Herd Instinct*, inasmuch as this title strongly suggests that he regards human societies as bovine or ovine or elephantine herds. But this is not really his intention; for he justly describes the German type of society before the war as "lupine." In contrast with it he says the British people were socialized like the bee—a comparison which anyone who reflects on the historical and geographical position of our nation can hardly hear of without a smile. Undoubtedly the British people were highly socialized; but their socialization may be easily interpreted (to adopt a word from the fashionable slang) as a "sublimation" of the hunting-life. For each of us was in eager pursuit of prey, symbolized by the sovereign (exchangeable for beef and mutton); we co-operated not too unselfishly toward that end; there were leaders and followers with much emulation; but some preparation had been made for the sick and injured; the attack was aggressive and carried out with strategy, energy and perseverance, not without jealousy between rival packs, and with occasional squabbles over the quarry.

CHAPTER VII

ACQUISITIONS IN CULTURE AND CUSTOM

§ 1. The differentiation of the human from the anthropoid stock must have begun a long time ago; as to when it began there is no direct evidence; and even if fossil remains of the earlier stages of our evolution had been discovered, we could only judge from the strata in which they occurred what must have been their relative antiquity. When it comes to reducing the chronology of past ages to figures, geologists either decline to make any estimate, or the results of their calculations may differ as 1 to 10. Since my own studies give me no claim to an opinion on such matters, whilst it is helpful to have clear ideas, however tentative, I have adopted the views of Sir Arthur Keith in his work on *The Antiquity of Man*, based on estimates published by Professor Sollas[1]. On turning to page 509 of that work, a genealogical tree will be found showing the probable lines of descent of the higher Primates. The separation of the human from the great anthropoid stock is represented as having happened at about the last third of the Oligocene period —say 2,000,000 years from the present time or (according to the later estimate, which I shall follow) 3,500,000—surely not too much to allow for our differentiation. Pithecanthropus (of Java) branched off as a distinct genus about the middle of the Miocene. Neanderthal man (*Homo Neanderthalensis*) and Piltdown man (*Eoanthropus Dawsoni*) separated as distinct species (or genera) from the stock of modern man (absurdly named *Homo sapiens*) early in the Pliocene, and became extinct respectively (say) 20,000 and 300,000 years ago. The races of modern man began to differentiate near the end of the Pliocene (say) 500,000 years from our own age. Such is the "working hypothesis."

[1] See the *Report of the British Association* (1900), pp. 711–30. The author has since then revised his estimates, assigning much greater depth to the Pliocene and Miocene deposits and proportionally more time for their formation. See the *Quarterly Journal of the Geological Society* (1909), LXV.

The skull capacity of the great anthropoids averages 500 c.c.; that of Pithecanthropus is estimated at 900 c.c.; the Australian native average is 1200 c.c.; Eoanthropus, according to Sir Arthur Keith, rises to 1400[1]; a Neanderthal skull has been measured at 1600 c.c.; the modern English average is under 1500 c.c. Of course, mental power depends not on the *size* only of the brain, but also on its differentiation, which may have been backward in *Neanderthalensis*, and in our own species may have advanced during the last three or four hundred thousand years.

§ 2. As to culture, the Neolithic period extends in Western Europe from about 2000 to 10,000 B.C.: and to that age is usually attributed the introduction of agriculture, the domestication of animals, pottery, weaving, permanent constructed dwellings, and monuments requiring collective labour; but some of these improvements, especially pottery and the domestication of animals, may be of earlier date. In other parts of the world, *e.g.* in the Eastern Mediterranean region, such culture is probably older but still comparatively recent. What is known as the Palæolithic stage of culture is often supposed to have begun early in the second quarter of the Pleistocene period, giving us a retrospect of (say) 300,000 years. But if we include under "Palæolithic" all unpolished stone-work that shows clear signs of having been executed according to an idea or mental pattern (and this seems a reasonable definition), the "rostro-carinate" implements must be so called, and then the beginning of this culture must be pushed back into the Pliocene[2]. In Pliocene and even in Miocene[3] deposits have further been discovered numerous "eoliths": stones so roughly chipped that they do not imply

[1] Dr Smith Woodward's reconstruction gives the skull of Eoanthropus a capacity of about 1300 c.c.

[2] See Ray Lankester's "Description of the Test-Specimen" (R.A.I., *Occasional Papers*, no. 4).

[3] Prof. Sollas has examined a collection of eoliths from upper Miocene deposits, and reports that we cannot, in the present state of our knowledge, understand how they can have been formed by natural agencies. "They seem to bear cogent evidence of design" (*Report of British Association* (1923), p. 475).

an idea-pattern; so that, whilst many archæologists accept them as of human workmanship, some experts dispute their claim to be considered artefacts. Of course, there must be eoliths; the only question is whether we have yet unearthed any of them. Our forefathers cannot have begun by shaping stones to a definite figure and special purpose. Beginning with stones taken up as they lay, they discovered that a broken stone with a sharp edge inflicted a worse wound than a whole one; then broke stones to obtain this advantage; used sharp fragments to weight clubs; and very slowly advanced to the manufacture of recognizable axes and spear-heads, meanwhile discovering other uses for flaked stones; and it seems to have needed at least 2,800,000 years to arrive at the poorest of known palæoliths. This strikingly agrees with the law, often stated, that the progress of culture is, by virtue of tradition, cumulative, and flows, as a stone falls, with accelerating velocity: in spite of the ebb, to which from age to age we see it to be liable. At any one time, moreover, the art of stone-working was, probably, even in adjacent tribes, at different stages of advancement. Indeed, we have no warrant to assume that every artefact discovered was once elaborated to the highest finish of which the workman was capable. 'Finish' depends partly upon the kind of stone obtainable. But it has been only very recently that such contrasts could occur as Herodotus[1] describes among the hosts of Xerxes: when, beside the well-accoutred Persians and Medes, marched Libyans and Mysians armed with wooden javelins hardened in the fire, and Ethiopians with stone-tipped arrows and spears headed with the sharpened horns of antelopes.

The moral of all this is (*a*) that there was abundant time before the rise of Neolithic culture (which may be called the beginning of civilization) for the complete adaptation of mankind everywhere, by natural selection, to the life of hunters; and (*b*) that, since then, there has not been time for the biological adaptation of any race to the civilized state. We shall see that natural selection has probably had some

[1] Book VII, chs. lxix, lxxi, lxxiv.

civilizing influence; but any approach to complete adaptation has been impossible, not only for want of time, but also because of rapid changes in the structure of civilization, the social protection of many eccentrics, the persistence of the hunting-life as a second resource or as a pastime, and by the frequent recurrence of warfare—that is to say, man-hunting. To civilization we are, for the most part, merely accommodated by experience, education, tradition and social pressure. A few people seem to be adapted to civilized life from their birth, and others to the slavish life; but all inherit, more or less manifestly, the nature of the hunter and warrior. This is a necessary basis of general and social psychology; and perhaps tribal or national characters (so far as distinguishable) may be understood by assigning the conditions under which they have, in various directions, been modified from this type.

To avoid the appearance of overlooking an obvious objection, I may add that the life of the hunter does not imply an exclusively carnivorous diet, but merely that hunting is the activity upon which his faculties are bent and upon which his livelihood chiefly depends. It is most unlikely that a cousin of the frugivorous anthropoids should entirely give up his ancestral food, immediately, or perhaps at any time. Even the diet of the wolf, in North-East Canada, includes "much fruit, especially the uva-ursi"; and the coyote there also eats berries[1]; so does the jackal in India. Savage women everywhere subsist largely on roots and fruits. Sir Arthur Keith says the teeth and jaws of the Neanderthal species were adapted to a coarse vegetable diet[2]. Yet the Neanderthal burials at La Ferrasie, La Chapelle aux Saints, Jersey and Krapina, with their implements and animal remains, leave no doubt that the species hunted the biggest game. At Krapina, besides the bones of mammoth and rhinoceros, those of "the cave-bear occurred abundantly, it was evidently a favourite article of diet"; the inhabitants were not fanatical vegetarians; and

[1] E. Thompson Seton, *The Arctic Prairies*, pp. 304, 352.
[2] *Op. cit.* pp. 151, 239, 476.

what a feat it was to slay with stone picks the cave-bear—bigger than the grizzly. They may have been trapped.

§ 3. Between the remote age when our hypothetical ancestor became a hunter and the time to which probably belong the remains of the oldest known men, there lies a gap of (say) two and a half million years, concerning which we have not only no direct evidence but not even any parallel in the world by means of which to apply the comparative method. Just at the beginning, the parallel of the wolf-pack sheds some light upon our path; but the light soon grows faint; for the primitive pre-human, from the first more intelligent than wolves, and inheriting from the ape-stock qualities of character which the new life greatly modified but could not extirpate, must under pressure of selection have become, after not many ages, an animal unlike any other. Just at the end, again, something concerning those who lived many thousand years before the beginning of history may be inferred from the parallel of existing savage customs; from their rock-dwellings, drawings, tools, weapons, hearths, something about their way of life; from evidence of their burial-customs, something of their beliefs. But what can be said of our ancestors during all those years that intervene between the beginning and the end?

Having been a hunter at the first and at the last, we may reasonably suppose that he had been so all the time. But, with our present knowledge, our chief guide as to other matters seems to be the fact that the most backward of existing savages possess powers of body and mind, and forms and products of culture, which must have been acquired gradually through a long course of development from no better origins than are traceable in apes and wolves. As the use of good stone weapons by living savages and the occurrence of stone weapons in deposits of various age in the Pleistocene—less and less perfectly made the further we go back—justify us in assuming that there must have been eoliths of even cruder workmanship at remoter dates, so the possession by savages of extensive languages, intricate customs, luxuriant myths, considerable reasoning powers and even humane sentiments, compel us to attribute such endowments to our prehistoric ancestors, in simpler and simpler forms, as we go back age by age toward the beginning. A tentative reconstruction of the lost series of events may sometimes be supported by what has been observed of the individual development of our children.

§ 4. For example, the constructive impulse, slightly shown by anthropoids that make beds and shelters in the trees, was called into activity in man especially in the making of weapons, tools and snares, and soon became an absorbing passion because he could not otherwise prosper; so that a savage (often accused of being incapable of prolonged attention!) will sit for days working at a spear or an axe: he is inattentive only to what does not interest him. Many children from about the sixth year come under the same sort of fascination—digging, building, making bows and arrows, boats and so forth. This is a necessary preparation for all the achievements of civilized life; and it is reasonable to suppose that the stages of growth of such interest in construction are indicated by the improvement of ancient implements.

§ 5. As to language—in the most general sense, as the communication of emotions and ideas by vocal sounds—the rudiments of it are widespread in animal life. A sort of dog-language is recognized, and monkeys seem to have a still larger "vocabulary." Hence, a number of emotional vocal expressions was probably in use among the primitive pre-human stock. And the new hunting-life was favourable to the development of communicative signs; for it depended on co-operation, which is wanting in ape-life, and in the lower extant savages hardly exists, except in hunting, war, and magical or religious rites. Hunting, moreover, is (as I have said) especially encouraging to onomatopœic expression in imitating the noises of animals and of weapons, and it was also favourable to the growth of gesture-language in imitating the behaviour of animals and the actions involved in circumventing and attacking them. Increasing powers of communication were extremely useful in planning and directing the hunt, and the pack must have tried to develop them. Without the endeavour to communicate, there could never have been a language better than the ape's; nor could there have been the endeavour without the need. That gesture alone was very helpful may be assumed; and it must have assisted in fixing the earliest vocal signs for things and actions and qualities, and probably determined the earliest syntax[1]; but when, in hunting, members of the pack were hidden from one another, or when their hands were occupied, gesture was not available, and communication depended on the voice. The speech of children similarly emerges from emotional noises and impulsive babbling, assisted by gesture. And we must observe that gesture (except the indicative) and onomatopœic noises have from the first a general meaning. To growl like a lion, signifies not a particular beast but any that makes a noise like that. Gesture and onomatopœia contain, therefore, the rudiment of those functions which make language an instrument of the most recondite thought. Meanings must have preceded the attempt to communicate; and the evolution of meanings is still in advance of language; and they (not words) constitute the materials with which thought operates.

Passing to later ages, we cannot expect to learn much about the speech of pre-historic man, whom we know only by a few bones. As to the Java skull, Sir A. Keith observes that "the region of the brain which subserves the essentially human gift of speech, was not ape-like in Pithecanthropus. The parts for speech are there; they are small, but clearly foreshadow the arrangement of convolutions seen in modern man." On the other hand, his "higher association areas...had not reached a human level[2]." The jaw of this skull not having been found, nothing can be said of its fitness for carrying out the process of articulation. As to Eoanthropus, "if our present conception

[1] For the syntax of gesture-language see Tylor's *Early History of Mankind*, ch. ii.

[2] *Antiquity of Man*, p. 268.

of the orbital part of the third frontal convolution is well founded, namely, that it takes part in the mechanism of speech, then we have grounds for believing that the Piltdown man had reached that point of brain-development when speech had become a possibility. When one looks at the lower jaw, however, and the projecting canine teeth, one hesitates to allow him more than a potential ability[1]." The jaw had not undergone the characteristic changes which in modern man give freedom to the tongue in the articulation of words[2]. But one "cannot detect any feature in the frontal, parietal or occipital areas which clearly separates this brain-cast from modern ones[3]." These areas include the "association centres," and their development indicates a great increase of intelligence. Eoanthropus, therefore, must have had a good deal to say and, being a social animal, must have felt the need of expression; and, though he was not a direct ancestor of ours, it can hardly be doubted that at some period the jaws of our own ancestors were no better adapted than his to articulate speech. May we not infer that articulate speech, meeting a need of the stock, arose very gradually, and was slowly differentiated from some less definite and structural connection of expressive and onomatopœic vocables, such as we have seen may naturally have arisen amongst the earliest hunters? *Pari passu* the jaw was modified.

§ 6. All savages live by custom; gregarious animals have their customs; and in the primitive hunting-pack customs must have been early established as "conditions of gregariousness." M. Salomon Reinach, indeed, thinks that the anthropoid probably became human as the result of inventing taboos, especially in sexual relations; there was then economy of nervous energy in the direction of the senses, and consequent enrichment of the intellect[4]. His hypothesis does not carry us far, perhaps, into the particulars of human form and faculty; but it contains this truth, that without the growth of customs

[1] *Op. cit.* p. 408.
[2] *Op. cit.* p. 452.
[3] *Op. cit.* p. 414.
[4] *Cultes, Mythes et Religions* III, p. 430.

there could have been no progress for human nature; and it certainly points to the probability that some custom was early established with regard to marriage. In Professor Westermarck's opinion our species was originally monogamous[1]. Supposing this to have been the custom, as it is amongst many Primates, could it have persisted after the formation of the hunting-pack? According to Mr Thompson Seton, wolves pair "probably for life[2]"; but this is disputed; and so it is whether or no the male of a seasonal pair takes part in caring for the puppies[3]. Of the primitive human stock one may say that whilst, on the one hand, the association of many males and females in the same pack may have tended to break up the family, on the other hand, the long youth of the children and the parental care generally characteristic of Primates would have tended to preserve it; that the practice of pairing requires the largest number of males (setting aside polyandry, which would have lowered the birthrate), and lessens quarrelling, and is therefore favourable to the strength of the pack; and that any custom may have been established that was most favourable to the species in its new life. The least probable of all conditions is promiscuity; for the rearing of children with their ever-lengthening youth must have been difficult, taxing the care of both parents. That we are nevertheless very imperfectly adapted to monogamy may be due to comparatively recent disturbances, such as the loss of seasonal marriage; but especially (I think) to the transition from the hunting-life to more settled conditions which required some sort of domestic labour; for this fell upon the women, and wives were multiplied for that purpose. The virtual enslavement of women captured in war, and the accumulation of women by chiefs and kings have corrupted and inflamed the sex-instinct with the pride of power—already tainted in that way.

§ 7. The claim to property is instinctive in most animals—claim to a certain territory, or to a nest, or lair, or mate. Each

[1] *History of Human Marriage*, ch. iii.

[2] *Life Histories of Northern Animals*, p. 757.

[3] See above, ch. v, § 4, note on p. 50.

early human pack probably claimed a certain hunting-range; and each family its lair, which it guarded, as our domestic dog guards the house. In Australia "every tribe has its own country, and its boundaries are well known; and they are respected by others[1]"; and the Boschmans, who retained the ancient hunting-life more perfectly than any other known people, are said to have been formerly divided into large tribes with well-defined hunting-grounds[2]. As weapons or other implements, charms, or ornaments came into use, the attitude toward the territory or lair will have been extended to include them; indeed, it seems to be instinctive even in lower Primates. "In the Zoological Gardens," says Darwin, "a monkey, which had weak teeth, used to break open nuts with a stone, and I was assured by the keepers that after using the stone, he hid it in the straw, and would not let any other monkey touch it. Here, then, we have the idea of property[3]." Among the half-wolf train-dogs of Canada, the claims of one to property seem to be recognized by others; for a dog will defend its *cache* of food against another that ordinarily it fears; and "the bigger dog rarely presses the point[4]."

The utility of keeping the peace within the tribe, no doubt, led to the growth of customs concerning property, and to their protection by the social sanction, and later by the taboo[5]. For taboo cannot be the origin of respect for property or for any custom: it implies a custom already existing, which it protects by the growth of a belief in some magical (or animistic) penalty that is effective even when there are no witnesses. The same utility of order must have established customs of dividing the kill of the pack: later also protected by taboo, as we still see in many savage tribes.

The attitude towards property is very variable amongst the tribes now known to us. Still, considering how early and strongly it is manifested by children, we may infer with some

[1] Spencer and Gillen, *Across Australia*, p. 198.
[2] G. W. Stone, *Native Races of South Africa*, p. 33.
[3] *Descent of Man*, ch. iii.
[4] E. Thompson Seton, *Life Histories of Northern Animals*, p. 769.
[5] E. Westermarck, *Origin and Development of Moral Ideas*, II, p. 52.

plausibility its antiquity in the race. The urgent desire of property, and tenacity in holding it, displayed by many individuals, though not an amiable, has been a highly useful trait, to which is due that accumulation of capital that has made possible the whole of our material and much of our spiritual civilization. Amongst barbarians it may be a necessary condition of social order. Had not wealth been highly prized amongst our own ancestors, it is hard to see how revenge could ever have been appeased by the wergeld. The payment, indeed, was not the whole transaction, it implied an acknowledgment of guilt and of the obligation to make amends; but these things would not have mollified an enemy nurtured in the tradition of the blood-feud, if silver had not been dear to him. It is still accepted as compensation for injuries that seem difficult to measure by the ounce. Wealth gives rank, and gratifies not only the greed but also the emulative spirit of the pack. Acquisitiveness is an essential trait of aristocracy, and adhesiveness of its perpetuity. Homespun prudence belongs, in our ancestry, to a more recent stratum of motives; we see it as a blind instinct in squirrels and beavers, a quasi-instinctive propensity in dogs and wolves (who hide food that they cannot immediately devour); but it is not known in any anthropoid, and is acquired at some stage by some human races—not by all; for it is not found in many extant savages. The only occasion on which Australian tribes show prudential foresight as to food is on the approach of the season of magical rites, when they lay in a stock of it, before giving themselves up for weeks or months body and soul to thaumaturgy[1].

Prudence is not, however, merely a function of foresight or intelligence, or else the Irish would be as prudent as the Scotch.

§ 8. The first wars, probably, were waged for hunting-grounds, and this may have been a revival, for the carnivorous anthropoid pack, of a state of affairs that existed amongst

[1] Spencer and Gillen, *Native Tribes of Northern Territory of Australia*, p. 27.

their ancestors at a much earlier date; for battles for a feeding-ground have been witnessed between troops of the lower Primates. Such a battle between two bands of langur (*Semnopithecus entellus*) has been described[1]; and Darwin relates after Brehm how "in Abyssinia, when baboons of one species (*C. gelada*) descend in troops from the mountains to plunder the fields, they sometimes encounter troops of another species (*C. hamadryas*), and then a fight ensues. The Geladas roll down great stones, which the Hamadryas try to avoid, and then both species, making a great uproar, rush furiously against each other[2]." As packs of the wolf-ape increased in numbers and spread over the world, they no doubt generally came to regard one another as rivals upon the same footing as the great cats and packs of dogs, and every attempt at expansion or migration provoked a battle. Wars strengthened the internal sympathies and loyalties of the pack or tribe and its external antipathies, and extended the range and influence of the more virile and capable tribes.

It is true that neighbouring tribes of savages are not now always mutually hostile. In Australia, we are told, local groups and adjacent tribes are usually friendly[3]; but with their sparse populations scattered (in most of the continent) over arid plains, the age of expansion seems to have closed some time ago, and a sort of equilibrium has been established. On the other hand, it is a shallow sort of profundity that insists upon interpreting every war as a struggle for nutrition, an effort to solve the social problem. Aggressiveness and insatiable greed are characteristic of many tribes—passions always easily exploited by their leaders, as in the civilized world by dynasts and demagogues. Plethora is more insolent than poverty. Lust of power, of glory, of mere fighting is a stronger incentive than solicitude for the poor.

However, in the development of society nothing has been so influential as war: an immense subject, for the outlines of which I refer to Herbert Spencer's ***Political Institutions***[4].

[1] *Royal Natural History*, I, pp. 72–3. [2] *Descent of Man*, ch. iii.

[3] Spencer and Gillen, *Across Australia*, p. 200.

[4] *Principles of Sociology*, II.

§ 9. Most of the amusements as well as the occupations of mankind depend for their zest upon the spirit of hunting and fighting, which they gratify and relieve, either directly or in a conventionalized and symbolical way, and which at the same time they keep alive. Sports and games involve the pursuit of some end by skill and strategy, often the seizing upon some sort of prey, or slaying outright, and they give scope to emulation. Emulation is a motive in the race for wealth, in every honourable career, even in addiction to science and learning; though here the main stress is upon an instinct older than the pack—curiosity, a general character of the Primates. That children at first play alone, later play together and then "make up sides," repeats the change from the comparatively solitary life of anthropoids to the social life and combined activities of the hunting-pack[1].

The chief interest of fiction (plays, novels, etc.), which makes such a large part of literature, lies in the plot, a series of incidents leading to some decisive event—renewing again and again the interest of the chase. The relation of means to an end as embodied in the chase, is such a fixed form of human thought that few are able to conceive of any natural product except as a work of design and as having a purpose. From the interest of the chase and the aggressiveness that is involved in it must also be derived all that we call "enterprise," whether beneficent or injurious: a trait, certainly, which there is little reason to regard as inherited from the anthropoid stock.

§ 10. The great amusement and pastime of feeding has, no doubt, descended to us in unbroken tradition, through harvest and vintage festivals, from the unbridled indulgence that followed a successful hunt. And I offer the conjecture that

[1] In contrast with this young orang-utans in their "talk" as well as "in their actions are the counterpart of human infants." Their cries and wheedling tones are adapted to the obtaining of help and affection, and they love their keeper. But on approaching puberty they lose all affection for him and become sullen and ferocious. Similarly, at six or seven years chimpanzees become dangerous. They are adapted to family life, but are not social animals. Lull, *op. cit.* p. 660.

the development of laughter and the enjoyment of "broad" humour (so often discussed) may be traced to these occasions of riotous exhilaration and licence. We may suppose, indeed, that these conditions began to prevail not in the earliest days of the ravenous pack, but after some advance had been made in the customs of eating. Savages usually cram to repletion when possible, and with huge gusto, for there may not soon be another opportunity. If uproarious feasting was advantageous physically and socially (as, till recently, we all thought it was), addiction to the practice was a ground of survival; and laughter (a discharge of undirected energy, as Spencer says[1]) being its natural expression and enhancement, shared in its perpetuation. This social foundation agrees with the infectiousness of laughter, with its connection with triumph and cruelty, and with the quality of the jokes that still throughout the world excite most merriment—practical jokes and allusions to drunkenness, the indecorous, the obscene. Sir Robert Walpole preferred such humour at his table as the most sociable; because in that everybody could take part. Many refinements have been introduced in polite circles; but it is in vain that one begins a theory of laughter with an analysis of the genius of Molière.

Similarly, I suppose that weeping, lamentation and the facial and bodily expression of grief were developed by the social utility of common mourning in tribal defeat and bereavement. In most cases of private grief there is a disposition to give way to it only in secrecy and seclusion; it is not well that the pack should know how much one is hurt and weakened. But in bereavement (which cannot be concealed) there are at the lower stages of culture customary forms of mourning, generally also under an animistic sanction (to appease the dead), which are irremissible. These forms, often mechanically carried out, have something of the exaggerated and self-immolative quality of mourning for public disasters. Survivals of them are still found amongst ourselves, and are most compulsive with the most backward minds.

[1] *Essay on Laughter*.

CHAPTER VIII

MORALIZATION OF THE HUNTERS

§ 1. As the earliest stages in the life of our race are unknown to us, we can only speculate as to the gradual growth of friendliness, compassion, magnanimity, general benevolence and other virtues. They cannot be explained merely by the hunting-life, which so easily accounts for greed, cruelty, pride and every sort of aggressiveness. Robert Hartmann writes: "It is well known that both rude and civilised peoples are capable of showing unspeakable and, as it is erroneously termed, inhuman cruelty towards each other. These acts of cruelty, murder and rapine are often the result of the inexorable logic of national characteristics and, unhappily, are truly human, since nothing like them can be traced in the animal world. It would, for instance, be a grave mistake to compare a tiger with a blood-thirsty executioner of the Reign of Terror, since the former only satisfies his natural appetite in preying upon other animals. The atrocities of the trials for witchcraft, the indiscriminate slaughter committed by negroes on the coast of Guinea, the sacrifice of human victims by the Khonds, the dismemberment of living men by the Battas, find no parallel in the habits of animals in their savage state. And such a comparison is, above all, impossible in the case of the anthropoids, which display no hostility toward men or other animals unless they are first attacked. In this respect the anthropoid ape stands upon a higher plane than many men." Are we, then, to explain the more amiable side of human nature, partly at least, by derivation from the frugivorous Primates, extensively modified by our wolfish adaptation, but surviving as latent character?

We ought not, I think, to attribute unnecessary cruelty or

destructiveness to Lycopithecus, except so far as the novelty of his life may have led him into excesses compared with the economical life of true, adapted carnivores. He too was a narrow-minded beast. All these atrocities cited by Hartmann (and the list might be extended into a volume) have resulted from the expansion of imagination; which, exaggerating in vague apprehension the evils of (say) witchcraft, conceives no limit to the punishment of it; or, again, under the power of superstition or inflamed nationalism, conceives no limit to the demands of the gods for sacrifice, or to the satisfaction of the pride of power. "I will pursue; I will overtake; I will divide the spoil; my lust shall be satisfied upon them." As to the mere violence of these fits of passion, there is no more devastating rage, when excited, than that of an orang or gorilla; but they must be excited by the aggression of others. Aggressive cruelty, no doubt, was born in the hunting-pack, but it did not attain the "truly human" dimensions until it was enlisted in the service of ideas, and then again (as we see in all persecutions and invasions) further inflamed by the contagious sympathy of the pack in full cry.

§ 2. No doubt the affections of family life continued from the anthropoid condition, in which the protracted youth of children had developed them even beyond the warm attachment which prevails among the lower Primates; and with the still more protracted youth of the stock, advancing toward the status of Man, the family affections became more enduring. The extension of goodwill beyond the family depends (*a*) upon occasions of co-operation and (*b*) upon the absence of occasions of rivalry and antagonism with those outside. The first of these conditions was daily supplied by the hunting life; but the second not until customs had been established of sharing the prey. Friendliness and the disposition to mutual aid must be so useful to a hunting-pack that is not merely seasonal but permanent (as I take ours to have been), both to individuals and to the pack as a whole, within certain limits (as that the wounded, sick or aged must not amount to an encumbrance), that we may suppose natural selection to have favoured the growth of effective sympathy, not merely in giving warning of danger or in mutual defence, but so far as it is actually found at present in backward tribes. It nowhere seems to be excessive; and its fuller manifestation in some civilized countries seems to depend less upon a positive increase of benevolence in the generality than on the breaking down here and there of conditions that elsewhere oppose and inhibit it. Thus the generosity, mercy and magnanimity that constitute the chivalrous ideal, seem to depend upon the attainment by a class (sometimes by individuals) of such undisputed superiority that there is no occasion for jealousy or rivalry in relation to others; for should the superiority be disputed, these virtues quickly disappear. Similarly what have been called the "slavish virtues" of charity, humility, long-suffering may arise amongst those who are free from rivalry because they have no hope of aggrandisement in wealth or honour, and who have indeed suffered long. In some fortunate social conjunctures these virtues may interfuse and permeate all classes; for they have a common root, and are active provided that circumstances do not inhibit them.

Ideas, again, are no less powerful for good than for evil. Charity and magnanimity have their exalted moods, which are communicated more or less widely and sometimes inspire the united action of great bodies of men; as may be seen in some religious or patriotic movements, never more conspicuously than in the abandonment of privileges in the French National Assembly on the 14th August. Moreover, since in individuals our complex nature varies in all directions, and amongst the rest in the direction of benevolence; and since any organ or quality that varies is apt to continue to do so, and may go on varying even beyond the limits of biological utility, why in human life may not this happen with benevolence (or with any other passion or virtue)? Hence in some men it expands with wonderful richness and beauty even to the sacrifice of themselves—nay, by excessive clemency or generosity, even to the injury of the tribe or of the race.

§ 3. The moral sense or conscience has been discussed by Darwin[1] "exclusively from the side of natural history"; so as that is the way of considering human nature in the present book, I shall epitomize his account of it; which, as far as it goes, seems to be true, though needing some explanation. He finds four chief conditions of the growth of a moral sense: (*a*) the social instincts lead an animal to take pleasure in the society of its fellows, to sympathize with them and to help them. (*b*) When the mind is highly developed, images of past actions and motives continually recur; "and that feeling of dissatisfaction or even misery which invariably results...from any unsatisfied instinct would arise as often as it was perceived that the enduring and always present social instinct had yielded to some other instinct, at the time stronger, but neither enduring in its nature nor leaving behind it a very vivid impression"—as, for example, hunger. (*c*) After language has been acquired, public opinion can be expressed, and becomes the paramount guide of action; though still "our regard for the approbation and disapprobation of our fellows depends on sympathy." (*d*) Social instinct, sympathy and our obedience to the judgment of the community are strengthened by the formation of habit. Darwin then proves successively these four positions.

Seeing the stress here laid upon sympathy, it may make the matter clearer if we observe that the word occurs in different senses. Under the first condition (*a*) that a social animal is led to sympathize with his fellows and to help them, we understand (1) that he shares in their pleasure or their pain (emotional sympathy) and (2) that he is moved to help them (effective sympathy): two states of mind that are by no means inseparable. Again under the third head (*c*) our regard for the approbation or disapprobation of our fellows is said to depend on sympathy: but knowledge of another's thoughts is with a social animal an automatic inference; it is not sympathy, except so far as, being accompanied with assent to his judgment, there is participation in his feelings of approval or

[1] *Descent of Man*, chs. iv, v.

disapproval; and if we dissent from his judgment, there may indeed be perceptive sympathy as to his feelings, but there is no participation in them; there is rather fear or resentment. One must bear in mind that (1) perception of another's feelings, (2) participation in them and (3) the impulse to help or relieve are separable processes, and that mere perceptive sympathy is as active in cruelty as in generosity or mercy. An animal that cannot perceive that it inflicts pain is incapable of cruelty.

It may be added that the second of the four conditions assigned by Darwin as determining the growth of the moral sense—reflective conflict between social instinct and the memory of an action opposed to it—accounts more especially for "remorse of conscience"; and that the third condition —the pressure of public opinion—explains that tone of "authority" attaching to conscience on which Bishop Butler laid so much stress[1].

§ 4. The chief addition which it seems necessary to make to the above considerations as to the growth of the moral sense consists in taking account of the influence of custom, the original foundation of both morals and law. All social animals have their customs, and Lycopithecus must early have acquired them, if the pack was to work together without fatal dissensions. There must have been customs of following leaders and of dividing the prey, and if the family maintained itself within the pack it must have been protected by custom. Edward Westermarck has shown that tribal custom was the first rule of duty, and that custom confers upon morality the three great qualities of generality, disinterestedness and impartiality[2]. Under custom everyone in the same circumstances does the same thing (it is general); the feeling finds place that "this is necessary," even apart from responsibility to anyone (it is disinterested); and if a custom is broken the trespasser (no matter who) is blamed, and he blames himself (it is impartial). This consideration greatly strengthens Darwin's argument.

[1] *Sermons on Human Nature.*
[2] *Origin and Development of Moral Ideas*, ch. v.

When a man (or animal capable of so much reflection) remembers an action of his own that was socially injurious, it is not merely the social instinct with which it is felt to be in conflict, but the definite custom which he knows that he has broken and which the social instinct supports. And, again, public opinion becomes the paramount guide of action not in relation to merely occasional lapses but as the sanction of custom, to the observance or breach of which approbation or disapprobation attaches.

The inadequacy of custom in relation to morality lies in its controlling only deeds (overt actions) and in necessarily neglecting the intention of the agent: a result of which is that still, in many backward societies, no account is taken of intention in awarding disapprobation or punishment. So far custom hinders moral reflection, which is entirely concerned with what the agent meant to do. But no doubt in the early stages of our development it was far more important that there should be definite customs known to everyone (though some of them may have been foolish enough) than that individuals should be left to their own private reflections upon duty whilst yet their reasoning powers and moral judgment were very feeble and confused. After a time—probably a very long time—some men suffering under disapprobation must have felt the difference it made in the situation according to whether they meant to offend or did not; they would next see that this distinction must hold good in the case of others, and if they had the courage to say so and influence to make themselves heard, moral judgment would become separable from customary judgment[1].

How early the moral sense began to form itself in our stock cannot be estimated because it must have been a very gradual process. Probably the rudiments of it appeared in the family life of the ape even before our differentiation; and the authoritative character of conscience established itself under the discipline of the hunting-pack before there was much development of mind (for dogs know what theft is) and under pressure

[1] I have discussed these matters in *Natural and Social Morals*, ch. vi.

of a public opinion which managed to express itself without articulate language. In his original and suggestive book Mr Trotter has shown that a herd (pack, tribe or nation) necessarily approves of whatever actions are done in its interests as good or right, and disapproves of the contrary actions as bad or wrong. Confident that its beliefs and customs are good and right, the pack persecutes dissenters and nonconformists. "Good" is a relative idea. "'The good are good warriors and hunters,' said a Pawnee chief; whereupon the author who mentions the saying remarks that this would also be the opinion of a wolf if he could express it[1]." Hence we may guess one of the principal contents of the primitive categorical imperative. The study of ethnology and history enables us to trace the modification and enrichment of these contents under varying conditions of culture, and for the results of such study I refer to Edward Westermarck's *Origin and Development of Moral Ideas*.

§ 5. After the introduction of agriculture (or rather of gardening) the stress of natural selection was in some directions altered. It was no longer confined to the perfecting of a society of hunters. At first, indeed, most garden work, probably, was done by women; but in its progress it fell extensively into the hands of men; and then advantage accrued to those tribes that were capable of steady industry and prudence. The new employment decreased aggression on the principle that "had Alexander been holding the plough, he could not have run his friend Clitus through with a spear." The sick and aged were now less of an encumbrance than they had been to hunters, and there was more scope for natural affection. Those who could not endure a settled life wandered away in their old pursuits. The more aggressive clans slaughtered one another in the vendetta. Social pressure and hanging (or its equivalent) eliminated many of the more idle, improvident, dishonest and unruly, whose instincts resisted "accommodation." The more neighbourly and co-operative tended to predominate. As

[1] Tylor, *Primitive Culture*, II, p. 89.

civilization intensifies, numerous ways of getting a livelihood, which (as we have seen) derive their motive force from the spirit of the pack, gratify that spirit under so many disguises and with so little direct personal collision, as to be compatible with a good deal of friendliness and benevolence; and co-operation direct or indirect steadily increases. But the primitive jealousy and grudging are not extinguished, and on the skirts of society prowls a ravenous pack, watching and gnashing its teeth.

Increasing capacity of forming ideas of remote ends and of co-ordinating various activities in their pursuit, implies the inhibition of many aggressive or distracting impulses that would frustrate our plans, and thereby constitutes an automatic control. And although it is now fashionable to depreciate the power of intelligence in human life, surely its development has had great influence. As men come to foresee the many consequences of an action, they learn to modify and regulate it, since each foreseen consequence, good or evil, excites some impulse either reinforcing or inhibiting the action. Reflections, too, upon all the circumstances of our lot has done much to ameliorate it. The "conditions of gregariousness" (to use W. K. Clifford's definition of morality) have been expounded by the more penetrating and comprehensive minds —prophets, poets and philosophers; and some disciples have understood them, and have persuaded many to believe. Nor have such luminaries arisen only in the later phases of culture when their writings have been delivered or their sayings recorded. Probably it was some one man who first had the insight and the courage to point out to a tribe that had ignored the fact, that whether a custom had been broken with a good intention or not affected the agent's guilt and ought to affect the penalty exacted. Some one man, probably, first saw what injustice is often disguised under the specious equality of the *lex talionis*; another first tried to assuage the bitterness of a vendetta and to restrain its destructiveness by appointing compensation; another, perhaps, first proposed to substitute animal for human sacrifice, or a puppet for a slave. And when we read the lists of sagacious proverbs that have been collected from many savage tribes, we must consider that it was by eminent individuals that those sayings were first uttered one by one: individuals with the gifts of insight and expression to summarize the experience of a whole tribe in memorable words, rude forerunners of the poets, philosophers and prophets.

CHAPTER IX

INFLUENCE OF THE IMAGINARY ENVIRONMENT

§ 1. It was set forth in Chap. V that the chief conditions to which mankind has been adapted and thereby differentiated in body and mind from the anthropoid stock are four—the hunting-life, geographical circumstances, social life and his own imaginations concerning Magic and Animism. This last condition remains to be discussed. It had, of course, nothing to do with the earlier stages of our evolution. Zoological man (if I may be allowed the expression) probably advanced to recognizable form and activity without the help of superstitions. But Man, as we know him, is everywhere a "political animal," one who realizes his nature only in society[1]; and societies are now everywhere held together by imaginations concerning occult powers. So far, therefore, as such beliefs have aided the preservation and development of society, and so far as Man as we know him owes his nature to such society, those beliefs have determined his development.

The necessity of learning the whole art of hunting from its rudiments without the help of instinct or tradition, by sheer observation, memory and inference, put extraordinary stress upon the brain. At first by knowledge, strategy, co-operation and persistence of will, later by devising weapons and snares, evolving language and discovering the ways of producing and utilizing fire, man found means of entirely changing the conditions of his life; but this would have been impossible without a great development of his brain; and, accordingly, it appears that Eoanthropus, at the beginning of the Pleistocene, had a skull with more than twice the cubic capacity of any anthropoid's. With the growth of the brain came a continually increasing fecundity of ideas: "Piltdown man saw, heard, felt, thought and dreamt much as we do[2]." The use of ideas is

[1] I suppose Aristotle meant "only in a City."
[2] Keith, *op. cit.* p. 429.

to foresee events and prepare for them beforehand: the great advantage of distance-senses over contact-senses is to give an animal time to adapt his actions to deferred events; and ideas give this power in a vastly higher degree. So far the utility of brains and ideas seems to be obvious. But in order that ideas may be useful in this way they must (one would suppose) represent and anticipate the actual course of events. If they falsely indicate the order of nature, or even represent to us beings and actions that do not exist at all, ideas may seem to be worse than useless.

Now when we turn to the lowest extant savages, they are found to possess, in comparison with apes, a considerable fecundity of ideas: constituting, on the one hand, a good stock of common sense, or knowledge of their country and of the properties and activities of the things and animals found there (including the ways of their fellowmen) and of how to deal with them, which enables them to carry on the affairs of a life much more complex and continuous than any animal's; but including, on the other hand, a strange collection of beliefs about magic and animism, which entirely misrepresent the course of nature and the effective population of the world. As to magic, they hold that they (or some of them) possess certain sticks or stones (talismans), or know certain words and invocations (spells) or certain ritual actions, which have a subtle virtue or invisible power by which it is possible to control the movements of game, increase the accuracy of their weapons, injure or slay their enemies at a distance, heal or resuscitate their friends, stimulate the fertility of plants and animals, direct the course of the weather and bring down the rain. As to animism (belief in the activity of spirits), they may be said to assume that whatever inexplicable event is not put down to magic must be due to the intervention of ghosts or spirits of some kind; though the action of spirits is often itself magical. Indeed, the only difference between the activity of magic and of spirits (both of them invisible powers) is that spirits are capricious, like animals and men, so that one cannot be sure how they will behave; whereas magical powers always act in the same way (as if according to laws of nature), unless they are counteracted by spirits or superior magic.

§ 2. These superstitions or imaginative delusions hamper savage men in so many ways, waste so much time and so much ingenuity in trying to control the invisible powers, so often restrict other more useful activities, lead them sometimes into such dark and cruel practices, that one might be excused for wondering whether their bigger brains can, on the whole, have been of any biological advantage to them in comparison with the anthropoids. The anthropoids live by common sense: so do savages, and they have much more of it; but the anthropoids seem not to be troubled about magic and animism. We must suppose that the common sense of primitive man (or of his forerunners) increased age by age as he became more and more perfectly adapted to the hunting-life, and that at some stage his imagination began to falsify the relations of things and the powers of nature. Primitive man is necessarily beset by anxieties and fears, anxiety about the success of his hunting, fears of enemies brute and human, especially at night; he is, therefore, eager to know and to control the future, and he seizes upon anything that seems a likely ground of anticipating and insuring future events. Within the narrow range of his daily life he has a practical sense of cause and effect, as in making tools and weapons, in lighting a fire, in hunting or fishing; but he has no analytic knowledge of the relations involved, and therefore cannot distinguish between causation and coincidence. This, I think, is the chief source of his particular magical beliefs—the mistaking of coincidental occurrences for the real causes of a well-directed action[1]. To us it might seem clear that the coincidental occurrence had no apparent connection with the event; but this difficulty is, for the savage, overcome by his belief in invisible forces that act at a distance. He has types of such forces in odours, radiant heat and sound; and there is the wind, an invisible force with palpable mechanical efficacy.

[1] See *Man and His Superstitions*, ch. ii, § 3, and ch. vi, § 5.

Such force he attributes to those circumstances that coincide with an event desired or feared, but whose connection with it cannot be perceived. Hence the world becomes full of invisible agencies that strike like spears, pierce like thorns, poison and heal like certain plants, and (amongst them) certain words and sentences that control events as a leader controls his followers (for there is some analogy for every delusion); and it seems to be the most important of all his concerns to defend himself against these powers or, if possible, to enlist them in his service. As to spirits and ghosts, the theory of Spencer and Tylor that the belief in them arises chiefly from dreams seems to me the most probable: no doubt illusions and hallucinations contribute something to the delusion; but, compared with dreams, these are rare occurrences. Savages in many cases believe that dreams are real objective experiences, and if in a dream they see a dead man they assume that he has really come to visit them[1]. Ghosts, therefore, are at first merely persons met and recognized under peculiar conditions, namely, when just waking from sleep and in spite of their being dead. They seem to be the type upon which all other spirits are imagined—indwelling powers of nature, demons, gods. We can only conjecture the stages through which what at first seemed a natural fact, though strange and unintelligible, the reappearance of the dead, became by the reporting of dreams or ghost-stories, and the construction of theories to account for them, spreading by general gossip and tradition throughout the community, another source of constant fear and anxiety, at first influencing funeral ceremonies and later giving rise to rites of sacrifice and propitiation or exorcism and finally of worship. Here it can only be said that these two groups of superstitions, Magic and Animism, become the dominant beliefs of savage life in relation to all affairs that are not comprehended in common-sense, the philosophies that explain all unfamiliar events and even some familiar enough (such as death) and the sanctions that control

[1] Spencer, *Princ. of Soc.*, §§ 49, 90; Tylor, *Primitive Culture*, chs. ix, x; *Man and His Superstitions*, ch. iii.

all conduct according to custom. I may add that the above explanations of these delusions, as having originated in coincidences and dreams, are by no means modern, but seem to be as old as Epicurus.

§ 3. At what stage of our history imagination, thus divorced from reality, began to influence human (or sub-human) life, it is impossible to say; but it cannot be less than half a million years ago, if Eoanthropus, 400,000 years ago, "thought and dreamt much as we do." Why did not such delusions hinder or entirely frustrate our development? Or did they promote it?

The first consideration is that biological adaptation is nearly always a compromise: if any organ or faculty be useful on the whole, in spite of some disutility, its increase favours the survival of those in whom it increases; and this is true of the brain and its thinking. The second is that superstitions, in spite of their wide prevalence and the fanaticism with which they are sometimes maintained, have not the stability of perception-beliefs founded upon the routine of daily experience: they remain plastic under the pressure of changing needs and modifiable from time to time by the demands of social utility: they are, in fact what Mr Canning Schiller calls "half-beliefs[1]." And a third consideration is that those magical and animistic delusions and practices that are socially destructive, probably belong to a stage of our life that was attained long after our differentiation from apedom had been established and when some progress had been made in arts and customs. Savages of the lowest culture have few beliefs that can be called positively injurious. When a hunter carries an enchanted spear and mutters a spell to control the quarry, but does not rely upon them without exercising all his acquired knowledge and skill, so that they are merely adscititious to common-sense actions, his magical precautions increase his confidence without weakening endeavour. To curse an enemy, or to "point the bone" at him, does not create but merely expresses a male-

[1] *Problems of Belief*, ch. v.

volent purpose; and, although sometimes fatal by suggestion, is on the whole better than to assassinate. Taboos do more good by protecting person and property and custom than they do harm by restricting the use of food. Belief in imaginary evils waiting upon secret sins exerts, whilst supported by social unanimity, a control upon all kinds of behaviour: it is the beginning of the "religious sanction" and one sort of conscience. The dread of spirits that prowl at night keeps people in the family cave or by the camp-fire; and that is the best place for them. Many rites and observances are sanitary. Totemism rarely does any harm, and may once have usefully symbolized the unity of social groups. Totemic and magical dances give excellent physical training, promote the spirit of co-operation, are a sort of drill; and (like all art), whilst indulging they also restrain imagination by imposing upon it definite forms. For a long time there was no special profession of wizard or priest, with whose appearance most of the evil of magic and animism originates; though perhaps even they generally do more good than harm by their courage and sagacity, by discovering drugs and poisons, by laying ghosts, and by their primitive studies in surgery, medicine and psychology.

§ 4. The wizard, however, and the priest, who could never have come into existence but for the prevalent beliefs in Magic and Animism (for of course they did not invent them), and who greatly extend and organize those beliefs, have a further and far more important function in human life, the organization or (rather) reorganization of human society. The organization of the hunting-pack described above was liable, as time went on, from several causes to fall asunder. Some of these causes are obvious: (*a*) Most important and earliest, I think, was the improvement of weapons and snares and the discovery of poisons which tended to make very small parties, or even single families, self-sufficing—as among the Boschmans (though a whole tribe sometimes assembled for a grand hunt[1]).

[1] G. M. Theal, *History and Ethnology of South Africa*, p. 11.

(*b*) Failure of game either from desiccation of the country, as in Australia (and as must have happened all along the great desert belt of the northern hemisphere), or else because the tribe had been driven into a poor country like Tierra del Fuego; so that a small population scattered over a wide area was reduced to a greater or less dependence on collecting. (*c*) The adoption of even a primitive agricultural or pastoral life may have made hunting a secondary interest. In such cases the natural leaders of a clan were no longer (as in the old pack) plainly indicated; and if society was to be saved from anarchy, some new form of control must establish itself for the preservation of tradition and custom. Possibly this may have happened in several ways; but in fact (I believe) we know of only one, namely, restraint by a supposed supernatural sanction: First, the rule of wizards, who are generally old men credited with mysterious powers that make the boldest clansman quail: such as the elders and headmen of Australian tribes. In New Guinea, too, and much of Melanesia the power of rulers, even though recognized as of noble birth, depends chiefly upon their reputation for magic. And among the Boschmans secrets about poisons and antidotes and colours for painting (probably considered magical) were heir-looms in certain families of chiefs and gave them caste[1]. Secondly, at a later stage, as the belief in ghosts more and more prevails, and ancestral ghosts are worshipped, and ghosts of heroes or chiefs become veritable gods, the priests who celebrate their worship strengthen the position of chiefs or kings descended from these gods, and help to maintain more comprehensive and coherent governments than those established upon magic only; though to these later forms, and to religions themselves, magic-beliefs make essential contributions.

If then these illusory imaginations or superstitions assisted early and also later culture, because they preserved order and cohesion by rearousing and directing the ancient submission and loyalty of the pack, it seems probable that the growth of superstition, or of some superstitions, may have been actually

[1] G. W. Stone, *Native Races of South Africa*, p. 76.

promoted by their utility. Still I agree with a friendly critic who urges that given the conditions, namely, man's primitive ignorance and the development of his imagination, the growth of this tangle of delusions was inevitable, even though they had all been useless or even injurious, as some of them were. If general and equally diffused, such beliefs could not have been specially injurious to any one tribe, since they would not have operated selectively, but merely as a burden and impediment to the whole race; whereas, if not equally diffused, any tribe that was lacking or relatively deficient in such beliefs, must have had an advantage, and superstition should then have declined by the elimination of the over-credulous. Indeed, however useful superstition may be in promoting tribal cohesion, it may in other ways often be excessive, amounting to a sort of tribal insanity and tending to destruction. Certain cases are well known, such as trial by the poison-ordeal, blind confidence in a wizard's predictions, taboos imposed as to food or sleep upon war-parties or hunting-parties, weakening them and impairing their efficiency, and so forth, that plainly tend to destroy or impoverish the tribe.

But my position is not that the utility of certain superstitions may be proved by general reasoning, but that their utility is shown to be highly probable by the fact that everywhere from the lowest savagery to modern "civilization," government and the preservation of custom and tradition have been closely associated with magical and animistic beliefs, and that the efficacy of such beliefs as sanctions of the conduct of the average tribesman or citizen has been generally acknowledged. That we have owed so much to delusions is a hateful thing to acknowledge: but I did not make the fact; I merely point to it: and we need not wonder that social life thus regulated should be so crude and embarrassed and disappointing.

But it is not easy to see how events could have been otherwise directed. The same mental development that enabled the growth of common-sense in primitive man also expanded his imagination. But common-sense is always limited to present conditions; it is nourished upon that which happens day by day; and its judgment depends upon the regularity of that sort of experience. It could never in a crisis of disorder, such as followed the incipient disorganization of the pack, have foreseen the dependence of human life upon order and the necessity of maintaining cohesion even at great immediate personal sacrifices; especially when we consider the low level of intelligence and character which had been an adaptation to life in the hunting-pack. These interests of order and cohesion were, however, as it happened, served indirectly, and without any foresight of the consequences, by the controlling power of certain delusions; and I think that natural selection must have favoured the survival of tribes whose members acquired an innate disposition to fall in with this sort of imaginative control.

§ 5. Perverse as it may seem, then, imaginations (most of them utterly false) have been an indispensable factor in promoting "progress"; but they themselves have been modified, and in some sort transformed, by adaptation to the mentality of progressive peoples. Thus have emerged from the lower animism the great religions in which the gods have been moralized and have become the guardians of morals and (seeing the heart) of conscientious behaviour and also of public law. By the studies and inquiries of priesthoods, again, necessary to their functions, the art of writing was discovered, history was recorded, and the sciences were established which were necessary to their worship (as of the sun) and to the building of temples and the perfecting of ritual—grammar, geometry, numbers, astronomy. In the building and services of temples and palaces, residences of ancient and present gods, the fine arts advanced and flourished—architecture, sculpture, painting, poetry, music and dancing—all tending to the great result of subduing mankind by the imagination. Meanwhile, agriculture and commerce, by the regularity of life they require, have, with the aid of records and the gradual diffusion of positive knowledge, greatly extended the province of common-sense, upon which we increasingly depend. The growth of social order and security diminishes the anxiety that was the great occasion of superstitions, and discourages the admission of that disorder in Nature which superstitions imply; whilst by making self-interest (not to boast of other motives) for most people a sufficient reason for observing the current morality, this social order greatly lessens the need of supernatural sanctions. But in the past at any rate the imaginary conditions of social life referred to in Chap. V, § 3, have by supporting government and civil order greatly helped in accommodating and even in some measure adapting us to our present condition, such as it is.

[Note to p. 5.] In the *China Journal of Science and Arts*, Jan. 1924, Prof. Osborne writes a short article on the results of the Third Asiatic Expedition of the American Museum of Natural History, in which he describes Mongolia as having been, since the middle period of the age of reptiles, "a vast, relatively high and fertile region" eminently suitable to the evolution of reptilian and later of mammalian life—"comparable in climate to the central plateau regions of Africa to-day." We may anticipate, he says, "that central Asia will be found to be the homeland of that highest division of the Primates which gave rise to man"—in some part of the plateau extending from eastern Mongolia across to western Turkestan; whilst the arboreal Primates lived in the low forest lands to the south. "The higher forms of ascent in the human family took place in open non-forested country, where bipedal rather than arboreal locomotion was established as the prevailing habit."

INDEX

Accommodation, 43
Adaptation, 8, 28, 42, 47, 68, 69
Aggressiveness, 57, 77
Agriculture, 2, 86, 97
Alimentary canal, 21
Amusements, 78
Animism, 48, 92; evil effects, 90; good effects, 94
Anthropoids, 14; habitat of, 26; mentality and habits, 31, 35, 44; skull-capacity 67; character, 80
Asia, Central, probable locus of Man's origin, 5
Atkinson, J. J., on *Primal Law*, 51
Australians, 23, 36, 67, 76, 77, 95
Avebury, on making fire, 35; baking clay, 37

Baboons, 18, 31, 77
Bears, 2
Belt, Th., on hairless dogs, 24; habits of monkeys sleeping, 24
Benevolence and friendliness, 81–82
Boschmans, 17, 75, 95
Brain development in Man, 21, 61
Brehm, on baboons, 77
Butler, J., on Conscience, 84

Cannibalism, 37, 58
Cats in the Oligocene, etc., 15
Chronology of geologic ages, 66
Clifford, W. K., 87
"Collecting stage," 46
Common sense, 90, 91
Comte, on early hunting life, 6 *n.*
Conscience, growth of, 83
Constructive impulse in anthropoid and Man, 44, 71
Co-operation in hunting, 32; the occasion of language, 33; amongst wolves, 59
Custom, 60, 73, 84–85

Darwin, C., on feet, hands, teeth and jaws of Man, 18, 21; sexual selection cause of naked skin, 22; vermin as possible cause, 23; early society, 51; growth of moral sense, 83
de Canteleu, on wolves, 50–56
Deniker, on races of Man, 29
Destructiveness, 55
Diet of Primates and hunters, 26
Differentiation of Man began in Oligocene primary difficulties, 3

Dogs, naked varieties, 24; primitive, 15; likeness to Man, 32

Emulation within the pack, 58; in business, games, etc. 78
Eoanthropus, 67, 89
Eoliths, 67
Erect gait in Man, 18

Families of different types, 49
Feet specialized, 19
Fire, how first made, 35
Freud, on *Totem und Tabu*, 52

Galton, on dog and man, 54; on moroseness of savages, 60
Garner, R. L., on speech of chimpanzee, 33; polygamy of apes, 45 *n.*
Generic consciousness of Primates, 42
Gesture, 72
Ghosts from ancestors, 92
Gibbon, 18, 31, 32
Glacial period, 27–28
Gregariousness, 56–63

Hands specialized, 19
Hartmann, R., on character of anthropoids, 80
Herd-life, 50; the character it developes, 63
Heredity, 1, 42
Herodotus on the army of Xerxes, 68
Hornaday, 4, 35 *n.*, 41 *n.*, 50 *n.*, 58
Hudson, W. H., on dogs, 59
Humanity, growth of the sentiments, Chap. VIII
Hunting life, influence of, on intelligence, 34, 61
Hunting-pack, its organization, 50, 53; organization lost, 94
Hypothesis, nature of, 8

Imagination-beliefs, 89
Intelligence of Man, how developed, 61; of anthropoids, 43; its influence on morals, 87

Keith, A., on length of intestine, 21; genealogical tree of Primates, 5; capacity of skulls, 67; Neanderthal species, 69; speech-organs of Pithecanthropus and Eoanthropus, 72
Köhler, Dr, on chimpanzees, 44

Language, 33, 71
Lankester (Sir R.), 67 *n.*
Laughter and lamentation, 79
Leader of pack, 58; of herd, 64
Livingstone, on S. African game, 17
Love of the chase, 54
Loyalty, 58
Lycopithecus, 3, 27, 32, 40; his stature, 14

MacBride, on specialization of Man, 6 *n.*
McCabe, 27
McDougall, on self-abasement, 58
Magic, 46; belief, how formed, 91; relation to government, 94–95
Man and His Superstitions, 48, 96–97
Marriage (seasonal), 38, 74
Maxim, H. S., on precedence, 59
Monkeys, speech of, 33; social, 31
Morris, Ch., Pref. and 7 *n.*, 27, 36

Natural and Social Morals, 84 *n.*
Natural Selection, 7–8; of best hands, 20; of the gregarious, 32; of the intelligent, 35; adapted Man to the hunting life not to civilization, 68; except partially, 88; favoured the superstitious, 97
Neanderthal Man, 26, 37, 67, 69
Neolithic culture, 67
Nordic race, 28

Oligocene, supposed age of differentiation of the *Hominidæ*, 5; landscape, fauna and flora, Chap. II
Onomatopœia, 34
Osborne (Prof.), 4, 11 *n.*, 98 *n.*

Palæolithic culture, 67
Parental care, 60
Pawnee chief, on the good man, 86
Pithecanthropus, 27, 67
Precedence in the hunting-pack, 58; amongst Australians, 59; in the herd, 64
Priest, 94
Primal Law, 51
Property, 58, 74, 75
Proteranthropus, 27
Psychology of the hunting-pack, Chap. VI
Pycraft, 51 *n.*
Pygmies, 27

Races of Man, 28–29, 66
Reinach, S., on sex tabu, 73
Religion, 97
Roosevelt, on dogs, 58, 60

Sanderson, G. P., on Indian wild dogs, 57
Schopenhauer, on female beauty, 23
Selous, F. C., on the hunter's joy, 54
Seton, Th., on Indians' love of killing, 55; on wolves, 56, 74
Sexual selection as cause of naked skin, 22
Shand, A. F., 46
Skin naked, Chap. III, § 5
Skull, changes in, 21; capacity of, 67
Social life of Primates, 31; of Man, 47–48
Sollas, on geological time, 5, 66; eoliths, 67 *n.*
Species of Man, 66
Speech in apes and Man, 33
Spencer, *Industrial Institutions*, 46 *n.*; *Political Institutions*, 77; on laughter, 79
Stages of culture, 46
Struggle over prey, 60
Superstition, utility of, 94, 98
Sympathy, perceptive, contagious, effective in the pack, 56; and moral sense, 83

Taboo, 75, 94
Territory claimed, 57–58
Topinard, 21
Totemism, 94
Trotter, on herd-instincts, 65, 86

Variability, 29

Wallace, A. R., on place of Man's origin, 6 *n.*; on human brain, 34; nakedness, 28; the orang, 62
War, 58
Weapons, 19, 62
Westermarck, on seasonal marriage, 38; primitive monogamy, 74; moral ideas, 84
Whewell, W., 8
Wizard, 94
Wolves, 55, 56, 59
Woodward (Dr Smith), 67 *n.*

Yerkes, R. M., on orangs, 43

For EU product safety concerns, contact us at Calle de José Abascal, 56–1°,
28003 Madrid, Spain or eugpsr@cambridge.org.

www.ingramcontent.com/pod-product-compliance
Ingram Content Group UK Ltd.
Pitfield, Milton Keynes, MK11 3LW, UK
UKHW040227290926
15471UKWH00001B/46
9781107628007